# Communications in Computer and Information Science 1709

More information about this series at https://link.springer.com/bookseries/7899

Tianrui Li · Rui Mao · Guoyin Wang · Fei Teng ·
Lei Zhang · Li Wang · Laizhong Cui ·
Jie Hu (Eds.)

# Big Data

10th CCF Conference, BigData 2022
Chengdu, China, November 18–20, 2022
Proceedings

Springer

*Editors*
Tianrui Li
Southwest Jiaotong University
Chengdu, China

Guoyin Wang
Chongqing University of Posts
and Telecommunications
Chongqing, China

Lei Zhang
Sichuan University
Chengdu, China

Laizhong Cui
Shenzhen University
Shenzhen, China

Rui Mao
Shenzhen University
Shenzhen, China

Fei Teng
Southwest Jiaotong University
Chengdu, China

Li Wang
Taiyuan University of Technology
Taiyuan, China

Jie Hu
Southwest Jiaotong University
Chengdu, China

ISSN 1865-0929          ISSN 1865-0937 (electronic)
Communications in Computer and Information Science
ISBN 978-981-19-8330-6          ISBN 978-981-19-8331-3 (eBook)
https://doi.org/10.1007/978-981-19-8331-3

This Springer imprint is published by the registered company Springer Nature Singapore Pte Ltd.
The registered company address is: 152 Beach Road, #21-01/04 Gateway East, Singapore 189721, Singapore

# Preface

Welcome to the proceedings of the 10th CCF Big Data Conference (BigData 2022), which was held in Chengdu, China, from November 18 to 20, 2022. Due to the COVID-19 pandemic and travel restrictions, the conference was held in a hybrid style (both on-line and on-site). The aim of BigData 2022 was to provide a high-quality platform for researchers and practitioners from academia, industry, and government to share their research results, technical innovations, and applications in the field of big data.

The topics of the accepted papers include theories and methods of data science, algorithms, and applications of big data. The papers were all comprehensively double-blind reviewed and evaluated by three to four qualified and experienced reviewers from relevant research fields. The eight full papers accepted for publication were selected from 28 submissions.

On behalf of the organizing committee, our thanks go to the keynote speakers for sharing their valuable insights with us and to the authors for contributing their work to this conference. We would like to express sincere thanks to CCF, the CCF Task Force on Big Data, Southwest Jiaotong University, Sichuan University, the University of Electronic Science and Technology, the Southwestern University of Finance and Economics, the Chengdu New Economic Development Commission, and the Sichuan Association of Artificial Intelligence for their support and sponsorship. We would also like to express our deepest appreciation to the Technical Program Committee members, reviewers, session chairs, and volunteers for their strong support in the preparation of this conference.

Last but not least, we highly appreciate Springer publishing the proceedings of BigData 2022.

November 2022

Tianrui Li
Rui Mao
Guoyin Wang

# Organization

## Honorary Chairs

Guojie Li      Institute of Computing Technology, Chinese Academy of Sciences, and Chinese Academy of Engineering, China

Hong Mei      Academy of Military Sciences, China

## Steering Committee Chair

Hong Mei      Academy of Military Sciences, China

## Conference Chairs

Weimin Zheng      Tsinghua University, China
Huan Liu      Arizona State University, USA
Dan Yang      Southwest Jiaotong University, China

## Program Committee Chairs

Tianrui Li      Southwest Jiaotong University, China
Rui Mao      Shenzhen University, China
Guoyin Wang      Chongqing University of Posts and Telecommunications, China

## Organizing Committee Chairs

Fei Teng      Southwest Jiaotong University, China
Lei Zhang      Sichuan University, China

## Conference Reward Chairs

Xiaoyang Wang      Fudan University, China
Yihua Huang      Nanjing University, China

## Publicity Chairs

Shaoliang Peng      Hunan University, China
Qian Yang      Golaxy, China
Xin Yang      Southwestern University of Finance and Economics, China

## Forum Chairs

Yan Yang                Southwest Jiaotong University, China
Ye Yuan                 Beijing Institute of Technology, China
Junming Shao            University of Electronic Science and Technology
                          of China, China

## Publication Chairs

Li Wang                 Taiyuan University of Technology, China
Laizhong Cui            Shenzhen University, China
Jie Hu                  Southwest Jiaotong University, China

## Sponsorship Chairs

Jidong Chen             Ant Group, China
Chang Tan               iFLYTEK, China

## Finance Chairs

Yuting Guo              CCF Task Force on Big Data
Chongshou Li            Southwest Jiaotong University, China

## Website Chairs

Chuan Luo               Sichuan University, China
Jin Gu                  Southwest Jiaotong University, China

## Program Committee

Alfred Liu              SAS, China
Anping Zeng             Yibin University of China, China
Bin Guo                 Northwestern Polytechnical University, China
Bo Jiang                IIE, CAS, China
Can Wang                Zhejiang University, China
Chao Gao                Southwest University, China
Chongshou Li            Southwest Jiaotong University, China
Chuan Luo               Sichuan University, China
Cuiping Li              Renmin University of China, China
Fengmao Lv              Southwest Jiaotong University, China
Gang Xiong              CASIA, China
Guanlin Chen            Zhejiang University City College, China
Guoyong Cai             Guilin University of Electronic Technology, China

| | |
|---|---|
| Haonan Luo | Nanjing University of Science and Technology, China |
| Hong Zhang | CNCERT, China |
| Hongjun Wang | Southwest Jiaotong University, China |
| Hongmei Chen | Southwest Jiaotong University, China |
| Hongmin Cai | South China University of Technology, China |
| Hongzhi Wang | Harbin Institute of Technology, China |
| Hu Ding | University of Science and Technology of China, China |
| Huanlai Xing | Southwest Jiaotong University, China |
| Ji Xu | Guizhou University, China |
| Jian Yin | Sun Yat-sen University, China |
| Jianyong Sun | Xi'an Jiaotong University, China |
| Jie Liu | Nankai University, China |
| Jieyue He | Southeast University, China |
| Jinpeng Chen | Beijing University of Posts and Telecommunications, China |
| Juanying Xie | Shaanxi Normal University, China |
| Jun He | Renmin University of China, China |
| Jun Long | Central South University, China |
| Jun Ma | Shandong University, China |
| Junchi Yan | Shanghai Jiao Tong University, China |
| Ke Li | Beijing Union University, China |
| Laizhong Cui | Shenzhen University, China |
| Lan Huang | Jilin University, China |
| Lei Zou | Peking University, China |
| Li Wang | Taiyuan University of Technology, China |
| Lin Shang | Nanjing University, China |
| Lizhen Cui | Shandong University, China |
| Lu Qin | University of Technology Sydney, Australia |
| Peiquan Jin | University of Science and Technology of China, China |
| Qi Liu | University of Science and Technology of China, China |
| Shengdong Du | Southwest Jiaotong University, China |
| Shichao Zhang | Central South University, China |
| Shifei Ding | China University of Mining and Technology, China |
| Shijun Liu | Shandong University, China |
| Tong Ruan | East China University of Science and Technology, China |
| Wangqun Lin | Beijing Institute of System Engineering, China |

| | |
|---|---|
| Wenji Mao | Institute of Automation, Chinese Academy of Sciences, China |
| Wu-Jun Li | Nanjing University, China |
| Xiang Ao | Institute of Computing Technology, CAS, China |
| Xiang Zhao | National University of Defense Technology, China |
| Xiaofei Zhu | Chongqing University of Technology, China |
| Xiaole Zhao | Southwest Jiaotong University, China |
| Xiaolong Zheng | Institute of Automation, Chinese Academy of Sciences, China |
| Xin Jin | Central University of Finance and Economics, China |
| Xin Yang | Southwestern University of Finance and Economics, China |
| Yanfeng Zhang | Northeastern University, China |
| Yanyong Huang | Southwestern University of Finance and Economics, China |
| Yi Du | Computer Network Information Center, Chinese Academy of Sciences, China |
| Yihua Huang | Nanjing University, China |
| Yilei Lu | Baihai Technology, China |
| Yiming Zhang | Xiamen University, China |
| Yuanqing Xia | Beijing Institute of Technology, China |
| Yubao Liu | Sun Yat-sen University, China |
| Yun Xiong | Fudan University, China |
| Yunquan Zhang | Institute of Computing Technology, Chinese Academy of Sciences, China |
| Zhaohui Peng | Shandong University, China |
| Zhen Liu | Beijing Jiaotong University, China |
| Zheng Lin | IIE, CAS, China |
| Zhenying He | Fudan University, China |
| Zhicheng Dou | Renmin University of China, China |
| Zhi-Jie Wang | Chongqing University, China |
| Zhipeng Gao | Beijing University of Posts and Telecommunications, China |
| Zhiyong Peng | Wuhan University, China |
| Zhonghai Wu | Peking University, China |

# Contents

# Searching Similar Trajectories Based on Shape

Zidan Fu and Kai Zheng[✉]

School of Computer Science and Engineering, University of Electronic Science
and Technology of China, Chengdu 611731, Sichuan, China
fuzidan@std.uestc.edu.cn, zhengkai@uestc.edu.cn

**Abstract.** Similarity search in moving object trajectories is a fundamental task in spatio-temporal data mining and analysis. Different from
conventional trajectory matching tasks, shape-based trajectory search
(STS) aims to find all trajectories that are similar in shape to the query
trajectory, which may be judged to be dissimilar based on their coordinates. STS can be useful in various real world applications, such as
geography discovery, animal migration, weather forecasting, autonomous
driving, etc. However, most of existing trajectory distance functions are
designed to compare location-based trajectories and few can be directly
applied to STS tasks. In order to match shape-based trajectories, we
first convert them to a rotation-and-translation-invariant form. Next, we
propose a distance function called shape-based distance (SBD) to calculate the accurate distance between two trajectories, which follows an
align-based paradigm. Then, to accelerate STS, we propose a trajectory
representation framework based on symbolic representation to support
efficient rough match. Finally, extensive experiments on two real-world
datasets demonstrate the effectiveness and efficiency of our framework.

**Keywords:** Trajectory data mining · Similarity search · Symbolic
representation · Trajectory compression

## 1 Introduction

Recent years has witnessed a tremendous growth in techniques and applications
based on location tracking devices, and massive trajectory data are generated
every day. There is a booming interest in processing and mining trajectory data
in large-scale datasets [1–3]. The technique of searching for trajectories that are
similar in shape can be applied to multiple scenarios. For example, in animal
movement analysis, we can infer animal species by their trajectories. For smart
vehicles, it is helpful to determine the proper route by the historical trajectories
which can make passengers feel natural and comfortable. And it is also useful
in recognizing geographical features of unfamiliar areas by analyzing historical
trajectories on it.

---

Z. Fu—Contributing authors.

T. Li et al. (Eds.): BigData 2022, CCIS 1709, pp. 1–20, 2022.
https://doi.org/10.1007/978-981-19-8331-3_1

There is a lot of work focused on this issue [4–8]. Traditionally, trajectory data is formalized by sequences of points sampled over a certain period of time, while the similarity of two trajectories can be ensured if they keep close to each other for the majority of their existence [9]. Unfortunately, these trajectory distance functions and pruning methods are mainly based on geographic proximity and can not be easily modified to determine the similarity between two trajectories with similar shape but in geographically remote places. The shape-based trajectory search (STS) problem faces three primary challenges:

*1) Transformation.* It is hard to calculate shape-based similarity of two trajectories by directly adopting traditional distance functions like Euclidean Distance (ED) [10] or Dynamic Time Warping (DTW) [3, 11, 12], unless an adequate transformation (translation and rotation) is introduced so that the compared trajectories can be spatially aligned, enabling the distance computed by these functions able to truly reflect the geometric similarity. When the data scale is large, the intensive and frequent calculation required for the transformation is unbearable. What's more, the state of "perfectly aligned" of two trajectories is hard to decide. Specifically, the distance to be translate and the angle to be rotated need to be resolved to ensure the distance of two transformed trajectories is minimum.

*2) Inconsistent Sampling Rate (ISR).* The sampling rate of real world trajectory data is not always fixed [9, 13]. The reasons vary but are mainly three folds. Firstly, due to the way data acquisition equipment work, some trajectory data is collected in a discrete manner at uncertain time intervals. Secondly, the speed of moving objects not only changes during one trip, but also varies in different trajectories. And this causes differences in distances between adjacent points in a trajectory with even stable sampling rate. Finally, a fraction of the sampled points are recorded incorrectly due to equipment errors or weak signals, which makes the problem even more complicated. With the sampling rate incomplete and imperfect in nature, we need to find a way to eliminate the errors caused by inconsistent sampling rates and ensure that the distance function can measure the true similarity among trajectories.

*3) Query Speed.* For real world application, query results need to be returned in a relatively short time, which requires a rather simple but effective method to organize data. For online scenes, efficiency of the algorithms are essential to guarantee the time consumption of massive query processes.

In this paper, a novel three-phase algorithm Align-based Trajectory Match (ATM) to calculate shape-based distance (SBD) is proposed. To address the challenges of transformation, we first transform the trajectory data into the angle/arc-length (AAL) space, and generate AAL sequences which is rotation and translation invariant. To overcome the problem of ISR, we develop a suite of algorithms to align the candidate trajectories to the query. The align algorithm try to find an optimum way to split and align the trajectories into pairs so that the distance of two trajectories can be computed by organizing the pairs. One of the advantage of this method is that it comprehensively considers the global matching effect of trajectories.

To accelerate the query process, we design an elastic symbolic representation scheme Symbolic Trajectory Match (STM). It converts the trajectory data to symbol series and only maintain the basic shape information of the trajectories. By comparing the symbol series of trajectories instead of the original trajectories, the customized framework can effectively filter the dissimilar trajectories out and speed up the query process.

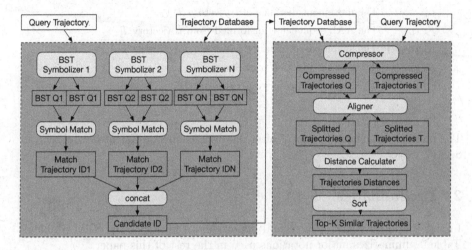

**Fig. 1.** Framework overview

Our contributions in this paper can be summarized as follows:

- We propose ATM, a three-phase framework to calculate the SBD of two trajectories. ATM employs a rotation-and-translation-invariant transformation and an align-based approach, which is robust under non-uniform sampling rates. Three alignment methods are designed to automatically find the optimum way of alignment. The computation of the SBD takes two kinds of distance into consideration (extra-distance and intra-distance), aiming to improve the overall precision and robustness of the algorithm.
- We propose a symbolic representation framework, BST, to store the summary of shape information of trajectories, which can be used to filter dissimilar trajectories efficiently. In addition, the establishment process of BST will not cause waste of computing power and storage, and can be applied both in batch-mode or online-mode.
- We conduct extensive experiments on real trajectory datasets and show the performance of proposed method.

The rest of the paper is organized as follows. In Sect. 3, the preliminary concepts are introduced and the problem is defined formally. Section 4 demonstrates algorithms to compute similarity functions of trajectories and Sect. 5 elaborates on the BST framework. Section 6 presents a brief introduction of related work.

**Table 1.** Notation

| Notation | Definition |
|----------|------------|
| $T$ | A trajectory |
| $\|T\|$ | The number of points in $T$ |
| $T[i,j]$ | Sub-trajectory of trajectory $T$ with the sequence from $P_i$ to $P_j$ |
| $TS_T[i,j]$ | Line segment generated by two points $P_i$ and $P_j$ of trajectory $T$ |
| $L_T$ | The movement distance of trajectory $T$ |
| $T_{AAL}$ | The AAL sequence transformed from trajectory $T$ |
| $SD(Q,T)$ | The shape-based distance between trajectory $Q$ and $T$ |
| $\overrightarrow{Ang_T}$ | The overall direction of trajectory $T$ |
| $S_T$ | The symbol trajectory of trajectory $T$ |

The paper is concluded in Sect. 7 after discussions on experimental validation in Sect. 5.

## 2   Preliminaries

In this section, we first give several definitions to formalize the STM problem. Table 1 summarizes major notations used in the rest of this paper.

**Definition 1** (Trajectory). A trajectory $T = [P_0, ..., P_n]$ is a temporally ordered sequence of $n+1$ points in two-dimensional plane, where $P_i = (x_{P,i}, y_{P,i})$ denote the position of the $ith$ point in $T$, and $|T| = n$ is the sequence length of trajectory. Besides, for any $0 \leq i \leq j \leq n$, $T[i,j] = [P_i, ...P_j]$ is a sub-trajectory of T.

Although the points are sampled in order, the time interval between consecutive points is highly uncertained. A detailed analysis of inconsistent sampling rate (ISR) has been given in [9] and it is demonstrated that ISR is ignored by many distance functions and may introduce errors to the result. To be mentioned, there are two types of ISR, namely time ISR and space ISR, which corresponds to different time intervals and different distance intervals between the adjacent points of the trajectory. For example, in the check-in dataset Foursquare, the intervals between two consecutive check-in record varies from seconds to hours. Even with precisely the same time interval all along the trajectory, distance intervals are almost all different due to different speed of moving objects. This uncertainty brings challenges to STS since it's hard to restore the original continuous trajectory with low sampling and sparse passing points [13]. In this paper, we ignore the velocity feature of the moving object and mainly analyze trajectories without temporal information. Thus, the similarity between two trajectories is completely determined by the shape of their underlying route. The STS problem of spatio-temporal trajectory is also an interesting topic but we will not discuss it in this research since the applicable scene may be quite different from what we are investigating.

With the location-based representation form of trajectories, a number of existing distance functions can be utilized in the computation of distance between two trajectories. However, little of them can be directly customized to solve the STS problem, in which the similarity of trajectories is not related to the distance of the trajectories.

Inspired by [14], a representation based on direction and length of trajectory segment is not affected by rotation and translation. Different from the way of representing trajectories by longitude and latitude, this form of trajectory consists of a series of end-to-end trajectory segments as the basic elements, which is more suitable to describe the shape information of the trajectory.

**Definition 2** (Trajectory Segment). A Trajectory Segment of a trajectory $T$ denoted as $TS(i,j) = \{P_i, P_j\}(i < j)$, is a line segment connecting two points $P_i$ and $P_j$.

With the definition of trajectory segment, a trajectory $T$ can also be represented as: $T_{lf} = [TS_T[0,1], ..., TS_T[n-1,n]]$.

**Definition 3** (Movement Distance). The movement distance of a trajectory or sub-trajectory $T$ denoted as:

$$L_T = \sum_{\forall TS_T[i,i+1] \in T} length(TS_T[i,i+1]) \tag{1}$$

is the sum of the length of each trajectory segment in $T$.

In addition to using the form of two endpoints, a trajectory segment can also be represented by a pair **(length, direction)**, which naturally discard its position information.

**Definition 4** (AAL Transformation). Given a trajectory $T = [P_0, ..., P_n]$, the corresponding format of $T$ after AAL transformation is:

$$T_{AAL} = [(Len_0, Ang_0), (Len_1, Ang_1), ..., (Len_{n-1}, Ang_{n-1})] \tag{2}$$

where $Len_i$ is the length of trajectory segment $TS(i, i+1)$, and $Ang_i$ is the angle between $TS(i, i+1)$ and $X$ axis.

In the rest of this paper, we use the terms "AAL sequence" or "AAL pair" interchangeably to refer to trajectory segment unless specified otherwise.

The direction attribute of trajectory segments can be defined in multiple ways, and the most widely used are absolute angle and relative angle. The absolute angle represents the angle between the trajectory segment and X-axis, and the relative angle represent the angle between a trajectory segment and its previous trajectory segment. [14] proved that the absolute angle is more accurate and robust than relative angle in similarity computation.

Given two AAL pairs, the distance between them can be naturally defined as the sum of the differentiation of length and direction attributes. However, these two attributes correlate to each other so tightly that the influence of them to the distance calculation follows a non-linear pattern. It is hard to tell which attribute is more important than the other, and when the length of the trajectory segment is large, a tiny change of the direction can make big difference to the trajectory.

Note that the two attributes together affect its shape geometrically, we adopt the modulus of difference of two vectors as the distance of AAL pairs.

**Definition 5** (Trajectory Segment Distance). Suppose $\vec{OA}$ and $\vec{OB}$ are two vectors generated from trajectory segments $TS1$ and $TS2$ respectively, the distance of $TS1 = [Len(TS1), Ang(TS1)]$ and $TS2 = [Len(TS2), Ang(TS2)]$ is defined as equation [TBD], which is the length of their endpoints.

$$dist(TS1, TS2) = \left\| \vec{AB} \right\| \tag{3}$$

[15] and [14] apply similar distance function based on AAL transformation to solve STM. Both use the length ratio $Len_i/L_T$ as the length attribute to keep the scale of trajectory invariant. For the distance computation, DTW and EDR technique are adopted to calculate trajectory similarity. However, both of these two schemes ignore the impact of inconsistent sampling rate, which can lead to a large number of false dismissals. The difference between our scheme and these two studies lies in the length attribute applied in the representation, and it can handle different sampling rates well. By using length instead of length ratio, trajectories in similar shape and proportion with query can be efficiently searched.

Now we formally define the problem investigated in our work.

*Problem 1.* Given a trajectory dataset $T$ and a querying trajectory $Q$, with a properly defined shape-based distance function, our goal is to find the trajectory $\tau \in T$ with the minimum value of $SBD(q, \tau)$, such that $\forall \tau' \in T \setminus \{\tau\}(SD(q, \tau) \leq SD(q, \tau'))$.

## 3   Align-Based Trajectory Match

In this part, we introduce a shape-based trajectory similarity algorithm, ATM, which mainly comprised of three parts: *1) trajectory preprocessing 2) trajectory segment matching 3) trajectory similarity calculation*

Note that the converted sequences are two-dimensional time series, two trajectories are similar if there is little difference between the corresponding trajectory segments of the trajectories. Intuitively, we can use distance functions such as Euclidean Distance, Dynamic Time Warping (DTW) [16] to match trajectories, which have been widely used in time series data. ED can be useful only when two trajectory sequences are in same length and well aligned. When the similar parts of the trajectories are not in the same position of the trajectory, it does not work. However, when the lengths of the sequences are not equal, or the sampling rates are not uniform, an external aligning method is necessary to map the trajectory segments from candidates to query. In [14,17], DTW is used to solve the STS problem by dynamically matching AAL pairs. It only works for continuous trajectories and fails when the scale of trajectories are considered. Besides, it ignores the fact that the AAL sequence is a kind of incremental sequence, which is not suitable for DTW when the sampling rate of the trajectories is inconsistent.

ATM is comprised of three phases: sequence normalization, sequence alignment and distance calculation. During the first phase, the direction attribute of candidate trajectory is normalized by rotating the entire trajectory by a proper angle, so that the global direction of the candidate trajectory is the same as that of the query. During the second phase, several methods are provided to align the trajectory segments of candidate to those of the query. During the third phase, the distance function are defined and used to calculate the overall distance between query and candidate.

## 3.1 Trajectory Preprocessing

Before applying the trajectory transformation, it is necessary to preprocess the trajectory first. On the one hand, it is necessary to reduce the complexity of the whole algorithm. On the other hand, it is necessary to erase the angle information of the trajectory, so that the processed trajectory is rotation-invariant.

### 3.1.1 Trajectory Compression

Since temporal information is not in consideration, merging consecutive trajectory segments in similar directions can bring little loss to the representation of the trajectories. Inspired by this, we can first compress the original trajectory based on shape. For convenience and simplicity, we use the Douglas Peucker (DP) algorithm to simplify trajectories. To be mentioned, any efficient and effective spatial trajectory compress algorithm can replace DP. By setting proper thresholds *tol*, we can ensure that the information loss is little and the simplified trajectory is basically in the similar shape of the original trajectory.

### 3.1.2 Sequence Normalization

In addition to compression, we obtain normalized trajectory samples by rotating the trajectory. Here we obtain a "trajectory trend" $\sigma$ by calculating the center of gravity of each trajectory, and rotate the entire trajectory according to $\sigma$.

After obtaining the AAL sequence of the candidate trajectory, normalization need to be done before distance calculation. As a two-dimensional time series, it is straightforward to normalize the length sequence and angular sequence separately. However, considering the correlation between length and angle attribute when determining the shape of the trajectory segment, as well as the cyclic characteristics of the angle, it is more reasonable to customize the normalization process. The normalization here aim to eliminate the difference in direction between the query and the candidate.

Then the overall direction of a trajectory can be defined by the average of all angles of trajectory segments in a trajectory. Note that the rotation angle may face problem of weak anti-noise ability, a more robust definition of overall direction based on vector computation is given.

**Definition 6** (Overall Direction). Given a data trajectory $T$ with start point $O$, $M$ feature points $P_1, ...P_i, ..., P_M$ are inserted to $T$ and satisfy that the movement length from $O$ to these points are $\frac{1}{M}L_T, ..., \frac{i}{M}L_T, ..., L_T$ respectively. Then the overall direction $Ang_T$ of trajectory $T$ is denoted as:

$$\overrightarrow{Ang_T} = \frac{\overrightarrow{F}}{|\overrightarrow{F}|} \qquad (4)$$

where $\overrightarrow{F} = \sum_{i=1}^{M} \overrightarrow{OP_i}$.

Given a query trajectory $Q$ and a candidate trajectory $T$, the offset which is obtained by calculating the difference between the overall directions of trajectories, is applied to all angle values in $T_{AAL}$. Thus the discrimination between the query and the candidate caused by rotation is largely eliminated.

## 3.2   Align-Based Trajectory Match

After preprocessing, trajectories are all rotated to the same direction. To calculate the similarity of two trajectories, we need to split them into the same number of trajectory segments and calculate the distance of segments individually.

By the way, there are some principles when designing an alignment algorithm:

- The basic unit of the algorithm is trajectory segments and the object is to find the optimal match of between two trajectories to minimize the global distance.
- An aligned pair of is a $M - to - N$ match, where $M$ and $N$ are independent and $M, N \geq 0$.
- The algorithm can handle the problem of ISR.

The align process can effectively solve the problem of ISR, error accumulation and make trajectories in different length comparable.

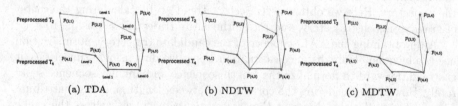

(a) TDA          (b) NDTW          (c) MDTW

**Fig. 2.** Align-based trajectory match

### 3.2.1 Top-Down Align Algorithm

The top-down align algorithm (TDA) is inspired by Douglas Peucker algorithm, which we used as a compression tool in trajectory preprocessing. It is based on alignments of points and follows a top-down paradigm. By searching the most representative point in a sequence and divide the sequence by this point recursively, the DP algorithm can ensure that the compressed trajectory preserve the key information of the original trajectory.

A natural thought here is that, if two trajectories are similar, the compressed skeleton trajectories of them and the aligned result of the compressed trajectory will be similar too and the splitting point generated by DP may be used to align two compressed trajectories.

When DP is used to compress trajectories, a top level point is selected and split the trajectory into two subtrajectories. The lower level splitting points in these subtrajectories split them into smaller ones. TDA record the order of the splitting points selected from DP with level ID until the algorithm finished. Then, in the aligning process, we merge the trajectory segments according to level ID.

### 3.2.2 Naive DTW for AAL Algorithm

Given the AAL representation of the trajectory dataset, DTW can be used to compute the trajectory similarities. Different from the traditional DTW, we use the trajectory segment distance defined in Eq. 3 to replace the point-wise distance.

However, the align result of NDTW is not ideal since it does not support many-to-many match. What's more, the one-to-many match in NDTW is also defective, because it only represent that one segment is similar to every single segment it is aligned to but not them as a whole. This property is useful to find similar trajectoris in scenery where stretch is not a sensitive factor, but is harmful to our problem.

### 3.2.3 Modified DTW for AAL Algorithm

In order to enable DTW to handle many-to-many match, we modify DTW to MDTW by updating the distance function to Eq. 5. For every step in dynamic programming, we try to find the optimal solution which support many-to-many match. Considering that the search space too large because the merging operation can cover the whole trajectory, the searching space is confined to $k$ steps.

Given two trajectories $T$, $Q$, the MDTW distance between $Q$ and $T$ is defined as:

$$d_{MDTW}(T,Q) =$$
$$\begin{cases} 0 & \text{if } |T| = |Q| = 0 \\ \infty & \text{else if } |T| = 0 \text{ or } |Q| = 0 \quad (5) \\ min\{dist(T[M-i:M], Q[N-j:N]))+ \\ d_{MDTW}(T[1:M-i], Q[1:N-j]),\} & \text{otherwise} \end{cases}$$

where $0 \leq i, j \leq k$ and $|T| = M, |Q| = N$.

*Example 1.* In Fig. 2, two trajectories compress from $T_2, T_4$ need to be aligned. First, the start points and the end points are aligned respectively. Next, the level 0 points $p_{(2,3)}, p_{(4,4)}$ are aligned. Then, the level 1 points $p_{(2,2)}, p_{(4,3)}$ are aligned. Finally, the level 2 points $p_{(4,2)}$ in $T_4$ are not aligned, because there is no level 2 points in the corresponding position in $T_2$. The align result of TDA is illustrated in Fig. 2a.

The align result of NDTW is illustrated in Fig. 2b. $T_2[1,2]$, $T_2[3,4]$ is aligned to $T_4[1,2]$, $T_4[4,5]$ respectively. Since $T_2[2,3]$ is similar to both $T_4[2,3]$ and $T_4[3,4]$, $T_2[2,3]$ is falsely aligned to $T_4[2,4]$.

Given $k = 2$, the align result of MDTW is illustrated in Fig. 2c. $T_2[1,2]$, is aligned to $T_4[1,2]$ and $T_2[3,4]$ is aligned to $T_4[4,5]$. Since $T_2[2,3]$ is similar to $TS_4[1,3]$, it is aligned to the combination of and $T_4[1,3]$.

### 3.3   Distance Calculation

Once the alignment of two trajectories are completed, the SBD can be easy to obtained. For one-to-many and many-to-many match, we need to update the distance function for aligned pairs. The overall distance contains two parts: external distance and internal distance. The external distance describe the difference of the aligned segments of two trajectories, while the internal distance of trajectory segment calculate the difference between it and its original subtrajectory.

**Definition 7.** Given two aligned trajectory segments $T[i,j]$, $Q[m,n]$ in $T, Q$ , the external distance and internal distance of the segments is defined as:

$$dist(T[i,j], Q[m,n]) = W_e * dist_e(T[i,j], Q[m,n]) + W_i * dist_i(T[i,j]) \tag{6}$$

$$dist_e(T[i,j], Q[m,n]) = dist(TS_T[i,j], TS_Q[m,n]) \tag{7}$$

$$dist_i(T[i,j]) = \sum_1^i d(P_{(T,i)}, TS[1,i]) \tag{8}$$

where $d(P, TS)$ is the vertical distance between point $P$ and line $TS$.

By summing up all the aligned pairs, the final SBD representation is obtained. Basically, more weights would be allocated to $W_e$, as the similarity of shape is more related to the overall trend of the trajectory. Moreover, a closer match of pairs can guarantee the relative position among all the segments. To be mentioned, TDA and NDTW need an extra process of distance calculation after alignment. while MDTW is able to dynamically calculate the distance in the process of alignment.

## 4   Symbolic Trajectory Match

Given a query, we can traverse all the trajectories in dataset to find the most similar trajectory using ATM, but the search time is not satisfying. No matter which align algorithm is chosen, the computational complexity is too high for large scale datasets or online applications. At the same time, storing the compressed skeleton points in advance will also bring a lot of burden to the system. If the compression ratio is too low, the performance of the query system would be undermined.

Inspired by some pioneer work in symbolic representation in sequential data, we propose Symbolic Trajectory Match system (STM), which consists of ATM and Multiple Balanced Symbol Trajectory (MBST). The MBST module maintains an approximate shape summary for each trajectory in dataset. It can support fast rough match and only require little space. In this research, MBST serves as a candidate filter for ATM. Besides, as a way of trajectory representation, it is potential in many applications such as trajectory compression, storage or data mining tasks [12].

In this section, an illustration of the method to generate shape-based trajectory symbols is given. Then we will discuss the distance of the symbol sequences. Finally, we demonstrate how to use STM to improve the performance.

### 4.1 Symbolic Representation for Trajectory

**Definition 8** (Symbol Trajectory). Given a trajectory $T$, its AAL representation $T_{AAL}$, and a mapping function $\{F|(Len, Ang) \leftrightarrow S\}$, the symbol trajectory of $T$ is denoted as:

$$S_T = [S_T[0], S_T[1]...S_T[n-1]] \tag{9}$$

where $S_T[i] = F(Len_i, Ang_i)(0 \leq i \leq z)$

This representation is generic for spatio-temporal trajectories but not convenient for STS since symbol trajectory is a mapping of a trajectory but not the underlying route. To ensure the uniqueness of the representation, Balanced Symbol Trajectory (BST) is proposed, which is constructed by a series of inserted points based on resampling techniques. With a parameter $L_{norm}$ confining the length of trajectory segments, the resampling process is functionally similar to the combination of normalization and alignment in Sect. 3.1. All the trajectory segments of the original trajectory are processed into a similar size for the convenience of the quantization process.

As a consequence, the mapping function can be simplified to $\{F'|Ang \leftrightarrow S\}$.

**Definition 9** (Balanced Trajectory). Given a normalized trajectory $T_n$ and a proper resample length $L_{norm}$, the corresponding Balanced Trajectory of trajectory $T_n$ is denoted as:

$$BT_T = [P_r[0], P_r[1], ..., P_r[i], ..., P_r[z]] \tag{10}$$

where $P_r[i](0 \leq i \leq z)$ is a resampled point of $T$ and the length of each trajectory segment $[P_r[i], P_r[i-1]]$ is equal to $L_{norm}$.

**Definition 10** (Balanced Symbol Trajectory). Given a Balanced Trajectory $BT$, its AAL representation $BT_{AAL}$, and a mapping function $\{F'|(Ang) \leftrightarrow S\}$, the corresponding Balanced Symbol Trajectory of trajectory $T$ is denoted as:

$$BS_T = [BS_T[0], BS_T[1]...BS_T[z-1]] \tag{11}$$

where $BS_T[i] = F'(Ang_i)(0 \leq i \leq z)$.

To construct the BT and BST, a trajectory resampling scheme and a quantization table which maps angles to symbols is needed. A lot of pioneer work make contributions to the problem of trajectory resampling, merging and recovering [9]. The selection of the algorithm needs to consider the fitness of the original trajectory and the execution speed of the algorithm. For quantization and mapping, it is important to decide the number of symbols in the mapping space and the mapping method is better to make the frequency of symbols in BST more evenly. As independant parts in the whole framework, they are quite convenient to be replaced by any other more advanced schemes. For simplicity, a linear interpolation scheme is used to generate BT as follows:

Comparing to symbol trajectory, there are two advantages using balanced symbol trajectory for STS task. On the one hand, BST is a one-dimensional sequence, so the computational complexity is obviously lower than that of ST, and the symbols of quantization are. On the other hand, BST can handle the ISR problem. If two trajectories are similar to each other, their ST representations may have much difference because the existence of ISR, but their BST representation would have much in common.

## 4.2   Symbol Distance Calculation

With the BST representation, the approximate distance of two trajectories can be calculated without access to the original trajectories. Since the number of the symbols is finite, we can calculate the distance between these symbols in advance and save the results in a lookup table to reduce the calculation cost. For example, with $L_{norm} = 1$, the corresponding lookup table of quantization map.

Since the trajectory segments in BT share the same length, the symbols in BST can be regarded as equally important and naturally aligned. Thus, the distance between two trajectories can be simply calculated by string matching algorithms like edit distances. For example, the distance of two BST $S_1 = $ "aab" and $S_2 = $ "aaf" can be calculated by looking up the table, so we can get $dist(S_1, S_2) = 0 + 0 + 2 = 2$.

## 4.3   Multiple Balanced Symbol Trajectory

In the resampling process, $L_{norm}$ is introduced and used as the length of all the normalized trajectory segments. With the value of $L_{norm}$ varying from big to small, the granularity described by the symbol sequences varies from coarse to fine. Larger $L_{norm}$ results in short sequences and may miss some important shape features of the original trajectory, while small $L_{norm}$ results in perfect description of trajectories and long redundant sequences with heavy calculation. We can make a compromise in choosing the value of $L_{norm}$ to achieve satisfying result. Another way to avoid the sacrifice for performance is a framework of Multiple Balanced Symbol Trajectory (MBST), which is constructed by multiple BST generated by different $L_{norm}$. By utilizing the match results of different BST system, the performance of MBST surpass that of single BST.

*Example 2.* Consider an example shown in Fig. 3, where $T_1, T_2, T_3$ are three data trajectories ready to generate their corresponding BST. In Fig. 3, given 3 data trajectories: $T_1, T_2, T_3$ and standard length $L_{n1}, L_{n2}$. Three trajectories are resampled by $L_1$ and $L_2$ separately. The BST of them under $L_1$ is: $[G, B, G, B, G, B, G, A], [H, H, H, G, H, B], [H, H, H, G, H, B]$. The BST of them under $L_2$ is: $[H, H, A], [A, H, A], [A, H, A]$.

## 4.4   Query Processing

With MBST fused with ATM, the framework adopts a filter-and-refinement approach. In particular, MBST works as a filter which takes relatively cheap computation cost to find a small set of candidate trajectories that are likely to be the results, which is a super set of the result for the original query. Then these

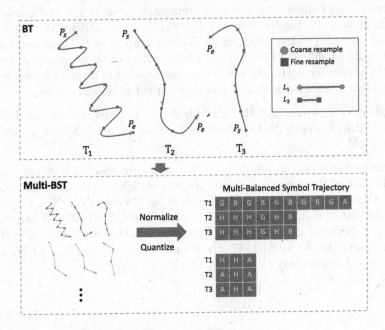

**Fig. 3.** Symbolic representation

candidate trajectories are further processed using ATM to obtain an actural result at the refinement step.

The whole framework is illustrated in Fig. 1. To be mentioned, all the BST of trajectories in database is calculated and stored. Subsequently, we discuss how to efficiently find the Top-K similar trajectories of query $Q$. At first, the MBST representation of query is calculated and compared to all the MBST in database. Then the match results of different BST is concatenated and the merged candidate trajectory ID is obtained and used to read original candidate trajectories from dataset. Finally, the distances between query and all the candidates are calculated and the Top-K similar trajectories are obtained after a sorting operation.

## 5    Experimental Study

We implemented our algorithms and conducted an extensive set of experimental studies, in order to verify the robustness of SBD, comparing different aligning methods and verify the pruning power of BST framework.

### 5.1    Experiment Setup

**Datasets.** We evaluate the performance of ATM and BST on two widely used datasets, i.e., Geolife[1] and ChengDu[2]. Didi contains trajectories generated by taxi drivers in Chengdu and the measurement interval of the track points is approximately 2–4 s. Geolife contains trajectories generated by 178 users in a period of over four years, which contains17621 trajectories. These trajectories were recorded by different GPS devices with a variety of sampling rates. We partitioned a trajectory in this dataset if the time interval between two consecutive point exceeds 15 min because they can be regarded as different trips.

**Evaluation Platform.** All the algorithms are implemented on an Ubuntu 16.04 operating system with 48-cores Intel(R) Xeon CPU E5-2650v4 @ 2.20 GHz 256 GB RAM.

**Parameter Setting.** In both datasets, we randomly selected 1000 trajectories and split each of them into two sibling trajectories with different sample points. Half of them are used as queries and others as ground truth. The trajectory movement distance varies from 5 km to 20 km. The default MDTW step is set to 4. The default value of $L_{norm}$ and direction resolution in BST is set to 600 m and 8 respectively. As for MDTW, the default value weights $W_e$ and $W_i$ is set to 0.8 and 0.2 respectively.

---

[1] http://research.microsoft.com/en-us/projects/geolife.
[2] https://gaia.didichuxing.com.

## 5.2   Experiment Results

### 5.2.1   Performance of ATM

We randomly select 100 trajectories from a dataset, and each trajectory is splitted into 2 sibling trajectories. One trajectory is used as query and the other is used as ground truth. Then we randomly select 99,900 trajectories from the rest of the dataset and construct data trajectories with the ground truth. For each query, we find the Top-K similar trajectories in the constructed dataset.

**Precision of SBD with Different Aligning Methods**

For precision, we calculate the MR, which is the mean rank of the sibling trajectories of each query, and precision, which counts the times when the sibling trajectory is calculated as the most similar one in test trajectories, with different ATM methods and the results is shown in Fig. 4. The precision represents how often the sibling trajectory is correctly measured as the Top-1 similar trajectory of query. The results clearly show that MDTW consistently outperform two other algorithms in terms of MR and precision.

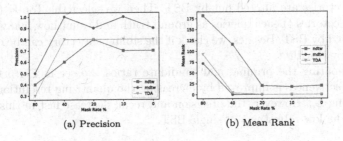

(a) Precision                    (b) Mean Rank

**Fig. 4.** Distance function precision evaluation

**Robustness of SBD with Different Aligning Methods**

For robustness, we randomly select 10,000 trajectories and construct two datasets, $D1$ and $D2$. While trajectories in $D1$ is complete, trajectories in $D2$ have some missing points. To quantify robustness, we first construct a ground truth result set by computing the Top-K similar trajectories list for each query from $D1$. Next, we recompute the Top-K list for the same query trajectory in $D2$. A robust distance function should overcome ISR and produce two similar answer sets on two datasets. Based on this hypotheses, we compute Spearman's rank correlation coefficient between the two Top-K lists. Since the elements in the two k-NN lists may not overlap completely, we form a single element set by taking their union. Finally, we compute the correlation between the two ranked lists. The closer the correlation is to 1, the more robust the distance metric is. All the three aligning algorithms are evaluated. The result in Fig. 5 shows that three algorithms all have good robustness and TDA outperforms MDTW and NDTW.

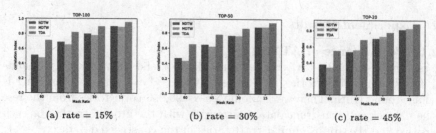

**Fig. 5.** Distance function robustness evaluation

### 5.2.2 Performance of BST

In this section, we evaluate the effectiveness and pruning power of BST. The pruning ratio is defined as follows:

$$Candidate\_ratio = \frac{|T_f|}{|T|} \tag{12}$$

Here, $T_f$ is the candidate set and $T$ is the trajectory set. We first find out how many trajectories are filtered out by BST. Then we select the Top-K ($k = 20$) similar trajectories of each queries as ground truth and count how many of them are preserved by BST. Besides, we check if the sibling trajectory can be correctly preserved by the filter.

By comparing the pruning and candidate ratios, we see in Fig. 6 that the pruning effectiveness is improved by increasing the quantizing resolution as well as decreasing $L_p$. However, the precision and recall decrease fast in this process and show the low robustness of single BST.

**Table 2.** Comparison of BST and Douglas Peucker

| Algorithm | Time (Seconds) | Storage (KB) |
| --- | --- | --- |
| Douglas Peucker | 309.01 | 2008.07 |
| BST | 29.79 | 70.71 |

BST is very efficient and powerful as a trajectory compression algorithm. We compare it with DP algorithm by compressing the trajectory data in Didi dataset. The size of the raw trajectory data is 58.5 MB. The result is shown in Table 2. As we see, BST is much faster than DP and the compressed file only occupies 70.71 KB.

Finally, the effectiveness and pruning power of MBST is evaluated. $M$ is set to 5 and use 5 different BST to search similar trajectories to queries. Then five sets of results are synthesized. The results are shown in Fig. 6g, h, i. As we can see, with the numbers of BST increasing, the precision remains relatively high. The top-20 recall decreases with high $M$ but it is acceptable considering the higher pruning efficiency. All in all, it is proved that more BSTs can make the framework more robust and efficient.

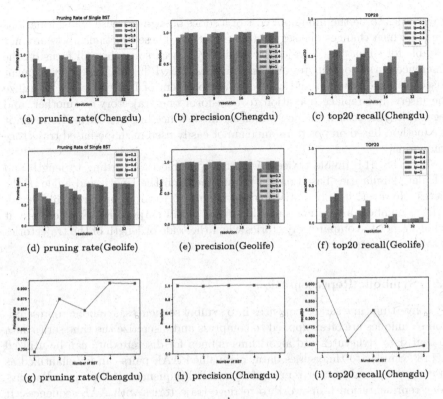

**Fig. 6.** Performance of BST

# 6   Related Work

An elaborate discussion on existing shape-based trajectory distance metrics and their weaknesses has already been done in Sect. 2. In this section, we introduce some important work in trajectory similarity measurement and symbolic representation.

## 6.1   Trajectory Similarity Measurement

Measuring the similarity between trajectories is a fundamental problem and has been studied extensively. Some classical solutions focus on indexing trajectories and performing similarity computation by the alignment of matching sample points.

Trajectory similarity measurement can be divided into two categories: point-wise functions and line-wise functions. As for the former, DTW is the first distance function to compute trajectory similarity and is designed to solve the time shift issue. EDR and ERP [16] are two edit-based functions influenced by the parameter setting heavily. On the other hand, some segment-matching methods are proposed. The major idea of EDS [6] is to define the distance between two trajectories to be the minimum cost of a series of segment-wise transformations each of which changes one segment to the other. These functions above are not suitable for shape-based trajectory similarity search since they all focus on the difference of the coordinates of points or lines. EDwP [9] is a trajectory segment based distance function and can solve the problem of ISR efficiently. It utilize the insert and replace operation to transform one trajectory to another, and regard the edit cost of the operations as the distance between them. However, it is a method based on positions and cannot easily used in shape-based trajectory comparing.

For STS, [14] divide trajectories into equal pieces according to length and calculate similarities based on DTW and focus on the length and angle of the pieces. However, it only works for continuous trajectories or curves like hand-written numbers and is not suitable for real world trajectories. Furthermore, it is not proper to compare trajectories when the scale or length of the trajectories is taken into consideration.

### 6.2  Symbolic Representation

By converting time series sequences into symbol sequences, symbolic representation techniques are often applied to compress and discretize the time series data so that data structures and algorithms defined for discrete data can be utilized [15, 18]. Similar to time series data, real-valued AAL pairs can be quantified as discrete pairs and further represent by symbol sequences. [15] proposed a symbolic representation framework to retrieve trajectories with AAL sequences. It first divide the length and direction of the trajectory into 8 equal parts individually and use 64 different symbols in total to represent any AAL pair. This symbolization method suffers from the ambiguity of the combination of consecutive trajectory segments.

## 7  Conclusion

In this paper, we studied the problem of matching trajectories based on the shape features which is rotation and translation invariant. We develop a suite of algorithms including an exact algorithm SBD and a hierarchical framework STM to address the uncertainty of the data trajectories caused by ISR. In addition, a symbolic representation framework STM was developed to prune the search space. At the same time, STM can be utilized to recognize similarities of trajectories in different granularity by customizing the parameters. Experimental study on real datasets verified the effectiveness and efficiency of our algorithms.

In the future work, we plan to discuss more types of similarity in shape. For example, similarities that are locally scale-invariant can be discussed and recognized to solve problems like gesture language comprehension. Besides, by employing data structure and algorithms for discrete techniques like hash, the power of symbolic representation can be further enhanced for extremely long and dense data trajectories.

# References

1. Wang, S., Bao, Z., Culpepper, J.S., Sellis, T., Qin, X.: Fast large-scale trajectory clustering. Proc. VLDB Endow. **13**(1), 29–42 (2019). https://doi.org/10.14778/3357377.3357380
2. Chen, Z., Shen, H.T., Zhou, X.: Discovering popular routes from trajectories. In: Proceedings - International Conference on Data Engineering, vol. 4(c), pp. 900–911 (2011). https://doi.org/10.1109/ICDE.2011.5767890
3. Sanchez, I., Aye, Z.M.M., Rubinstein, B.I.P., Ramamohanarao, K.: Fast trajectory clustering using Hashing methods. In: Proceedings of the International Joint Conference on Neural Networks, pp. 3689–3696 (2016). https://doi.org/10.1109/IJCNN.2016.7727674
4. Astefanoaei, M., Cesaretti, P., Katsikouli, P., Goswami, M., Sarkar, R.: Multi-resolution sketches and locality sensitive hashing for fast trajectory processing. In: GIS: Proceedings of the ACM International Symposium on Advances in Geographic Information Systems, pp. 279–288 (2018). https://doi.org/10.1145/3274895.3274943
5. Li, X., Zhao, K., Cong, G., Jensen, C.S., Wei, W.: Deep representation learning for trajectory similarity computation. In: Proceedings - IEEE 34th International Conference on Data Engineering, ICDE 2018, pp. 617–628 (2018). https://doi.org/10.1109/ICDE.2018.00062
6. Xie, M.: EDS: a segment-based distance measure for sub-trajectory similarity search. In: Proceedings of the ACM SIGMOD International Conference on Management of Data, pp. 1609–1610 (2014). https://doi.org/10.1145/2588555.2612665
7. Yao, D., Cong, G., Zhang, C., Bi, J.: Computing trajectory similarity in linear time: a generic seed-guided neural metric learning approach. In: Proceedings - International Conference on Data Engineering, pp. 1358–1369 (2019). https://doi.org/10.1109/ICDE.2019.00123
8. Frentzos, E., Gratsias, K., Theodoridis, Y.: Index-based most similar trajectory search. In: Proceedings - International Conference on Data Engineering, pp. 816–825 (2007). https://doi.org/10.1109/ICDE.2007.367927
9. Ranu, S., Deepak, P., Telang, A.D., Deshpande, P., Raghavan, S.: Indexing and matching trajectories under inconsistent sampling rates. In: Proceedings - International Conference on Data Engineering, pp. 999–1010 (2015). https://doi.org/10.1109/ICDE.2015.7113351
10. Yanagisawa, Y., Akahani, J., Satoh, T.: Shape-based similarity query for trajectory of mobile objects. In: Chen, M.-S., Chrysanthis, P.K., Sloman, M., Zaslavsky, A. (eds.) MDM 2003. LNCS, vol. 2574, pp. 63–77. Springer, Heidelberg (2003). https://doi.org/10.1007/3-540-36389-0_5
11. Yu, C., Luo, L., Chan, L.L.H., Rakthanmanon, T., Nutanong, S.: A fast LSH-based similarity search method for multivariate time series. Inf. Sci. **476**, 337–356 (2019). https://doi.org/10.1016/j.ins.2018.10.026
12. Driemel, A., Silvestri, F.: Locality-sensitive hashing of curves. In: Leibniz International Proceedings in Informatics. LIPIcs, vol. 77, no. 614331, pp. 371–3716 (2017). https://arxiv.org/abs/1703.04040. https://doi.org/10.4230/LIPIcs.SoCG.2017.37
13. Zheng, K., Zheng, Y., Xie, X., Zhou, X.: Reducing uncertainty of low-sampling-rate trajectories. In: Proceedings - International Conference on Data Engineering, pp. 1144–1155 (2012). https://doi.org/10.1109/ICDE.2012.42

14. Vlachos, M., Gunopulos, D., Das, G.: Rotation invariant distance measures for trajectories. In: KDD-2004 - Proceedings of the Tenth ACM SIGKDD International Conference on Knowledge Discovery and Data Mining, pp. 707–712 (2004). https://doi.org/10.1145/1014052.1014144

15. Chen, L., Özsu, M.T., Oria, V.: Symbolic representation and retrieval of moving object trajectories. In: MIR 2004 - Proceedings of the 6th ACM SIGMM International Workshop on Multimedia Information Retrieval, pp. 227–234 (2004). https://doi.org/10.1145/1026711.1026749

16. Chen, L., Özsu, M.T., Oria, V.: Robust and fast similarity search for moving object trajectories. In: Proceedings of the ACM SIGMOD International Conference on Management of Data, pp. 491–502 (2005). https://doi.org/10.1145/1066157.1066213

17. Vlachos, M., Kollios, G., Gunopulos, D.: Discovering similar multidimensional trajectories. In: Proceedings - International Conference on Data Engineering, pp. 673–684 (2002). https://doi.org/10.1109/ICDE.2002.994784

18. Lin, J., Keogh, E., Wei, L., Lonardi, S.: Experiencing SAX: a novel symbolic representation of time series. Data Min. Knowl. Disc. **15**(2), 107–144 (2007). https://doi.org/10.1007/s10618-007-0064-z

19. Lin, J., Khade, R., Li, Y.: Rotation-invariant similarity in time series using bag-of-patterns representation. J. Intell. Inf. Syst. **39**(2), 287–315 (2012). https://doi.org/10.1007/s10844-012-0196-5

# Unsupervised Discovery of Disentangled Interpretable Directions for Layer-Wise GAN

Haotian Hu[1], Bin Jiang[1,2(✉)], Xinjiao Zhou[1], Xiaofei Huo[1], and Bolin Zhang[1]

[1] College of Computer Science and Electronic Engineering, Hunan University, Changsha 410082, Hunan, China
{huhaotian,jiangbin,zhouxinjiao,hxfhnu,onlyou}@hnu.edu.cn
[2] Key Laboratory for Embedded and Network Computing of Hunan Province, Hunan University, Changsha 410082, Hunan, China

**Abstract.** Many studies have shown that generative adversarial networks (GANs) can discover semantics at various levels of abstraction, yet GANs do not provide an intuitive way to show how they understand and control semantics. In order to identify interpretable directions in GAN's latent space, both supervised and unsupervised approaches have been proposed. But the supervised methods can only find the directions consistent with the supervised conditions. However, many current unsupervised methods are hampered by varying degrees of semantic property disentanglement. This paper proposes an unsupervised method with a layer-wise design. The model embeds subspace in each generator layer to capture the disentangled interpretable semantics in GAN. And the research also introduces a latent mapping network to map the inputs to an intermediate latent space with rich disentangled semantics. Additionally, the paper applies an Orthogonal Jacobian regularization to the model to impose constraints on the overall input, further enhancing disentanglement. Experiments demonstrate the method's applicability in the human face, anime face, and scene datasets and its efficacy in finding interpretable directions. Compared with existing unsupervised methods in both qualitative and quantitative aspects, this study proposed method achieves excellent improvement in the disentanglement effect.

**Keywords:** Discovery of interpretable directions · Generative adversarial network · Unsupervised learning · Disentangled semantic

## 1 Introduction

Powerful image synthesis abilities and the capacity to fit domain-specific semantic information from enormous volumes of data [1–4] are two features of Generative Adversarial Networks (GANs) [5]. However, GANs do not offer a simple explanation of how it understands and utilizes the learned semantics. Until

H. Hu, X. Zhou, X. Huo and B. Zhang—Contributing authors.

**Fig. 1.** Examples of interpretable directions we found on the FFHQ and Anime face datasets. For example, the "Age" attribute and "Smile" are found in layer 1 and layer 4 of the face dataset, the "Pose" is found in layer 2 of the Anime face dataset and "hairstyle" in layer 3 of the Anime face dataset.

recently, [6] analyses pre-trained convolutional neural networks and generative adversarial networks by introducing network dissection [7,8]. They find that different layers in the CNN and GAN contained units corresponding to various High-level visual concepts that are not explicitly labeled in the training data. Many findings [8–11] also show that different layers in the network can capture semantic objects with different levels of abstraction. For instance, [12] analyzes the semantic information in the scene dataset using a pre-trained GAN model. They conclude from the experimental data that the model's deep, intermediate and shallow networks correspond to color background information, entity objects, and scene structure. In order to identify semantic properties from the different layers of the generator and to be able to control them to synthesize images, many studies [3,13–24] have focused on mining the semantic information in the latent space of the GAN in recent years.

Some methods [3,13–21] add supervised conditions to the learning of GAN and discover semantic directions in latent space consistent with supervised factors. [3,13–17,19–21] are used to add supervised conditions by assigning labels, manual annotations, or pre-trained classifiers to the generated images, thus finding interpretable semantic directions. For example, the approach [15–17] draw on the contrastive pre-training language-image encoder [18] and textual information as supervised conditions to guide the generation of semantic images. Recent works, such as [15,19–21], employ pre-trained attribute classifiers as supervised conditions to steer the semantic direction in the latent space of the GAN to be consistent with the specific attribute operations. However, supervised methods are limited to finding directions that are interpretable in light of the given supervision criteria; they cannot find a wide variety of semantic directions.

Another research direction that finds interpretable semantics in latent space is to impose unsupervised constraints on the orientation of the latent space. GANSpace [22] discovers important directions in GAN latent space by applying PCA in StyleGAN latent space and BigGAN feature space. SeFa [23] discovers GAN learned latent semantic directions by decomposing model weights, and it focuses on the relationship between image changes and internal representations.

However, these methods require heavy training efforts, and they need to randomly extract a large number of random latent directions and fit them to interpretable semantic directions as much as possible. EigenGAN [24] embeds a linear subspace in each generator layer. The orthogonal basis in subspaces at various levels can capture different semantic directions during model training. However, EigenGAN only imposes a simple regularization constraint on the subspace, and its network structure is relatively simple. For these reasons, although EigenGAN can unsupervised discover many interpretable directions, these directions are poorly disentangled, i.e., there are often multiple attributes entangled together. Compared with supervised methods, unsupervised methods are able to discover more interpretable directions than expected. In fact, poor disentangled visual effects are still a challenge that many unsupervised methods face.

This work aims to develop a network to explore more interpretable semantics in GAN's latent space. So we design a network structure with a linear subspace [24] in each generator layer to discover highly disentangled interpretable directions through an unsupervised approach during GAN learning semantic knowledge (Fig. 1 shows the examples). However, unsupervised approaches are usually worse than supervised ones regarding the degree of disentanglement of the found semantic directions. Inspired by [3], we introduce the mapping network of styleGAN to improve the disentanglement of interpretable directions discovered unsupervised. Many studies [3,25,26] also show that the latent space $W$ of styleGAN is rich in disentangled properties and that the space $W$ can learn the more disentangled semantic information better than the original space $Z$. Inspired by this, we introduce the latent mapping network in the network structure in order to improve the disentanglement of attributes of interpretable directions discovered in an unsupervised way. In addition, to further improve the disentanglement between the subspace feature dimensions, we introduce Orthogonal Jacobian regularization [27] to disentangle the model by constraining the orthogonal properties between changes caused by the output of each feature dimension. Compared with the method of EigenGAN, we directly impose constraints on the model's inputs to disentangle the learned directions of feature dimensions in each layer. The trial outcomes demonstrate a significant improvement in disentanglement for our strategy.

Overall, we would like to emphasize the following as our primary contributions:

- We suggest an unsupervised method for discovering disentangled interpretable directions in a layer-wise GAN's latent space.
- To overcome the attribute entanglement problem of unsupervised methods, we add a latent mapping network and Orthogonal Jacobian regularization to the model. The latent mapping network transforms the generator's input to an intermediate latent space with rich disentangled semantics. Meanwhile, Orthogonal Jacobian regularization imposes constraints on the overall w-vector of the input generator to improve its orthogonality.
- The experiment results demonstrate our approach's ability to identify distinct and disentangled semantics on various datasets (e.g., human face, anime face,

scene). Compared with existing unsupervised methods in both qualitative and quantitative aspects, ours achieves excellent improvement in the disentanglement effect.

## 2 Related Works

### 2.1 Generative Adversarial Networks

GAN's fundamental structural components are a generative network and a discriminative network. The goal of the generative network is to map the noise obtained from random sampling to a high-fidelity image while fooling the discriminative network as much as possible. On the other hand, the discriminant network must decide if the image sent by the generative network is genuine or fake. Therefore, the training is completed by gradually making the generative network capable of generating realistic images as the generative network and the discriminative network confront each other.

Because the training process of GAN is usually complex and unstable, plenty of researches carry out on regularization [1,28,29] and loss function [30,31] in order to improve the ability of GAN to learn semantic knowledge. Our proposed method requires designing subspace structures in each layer of the generative network such that it can sense the changes in sample distribution from random noise fitting to generate realistic images and capture interpretable changes as semantic directions.

### 2.2 Semantic Discovery for GANs

After GAN models were developed, it was discovered that the latent space of GAN typically contains semantically significant vector operations. Therefore, many studies [8,12–14,22,23,32–37] have been devoted to mining these vectors and using them for image editing.

**Supervised Methods.** For some methods [8,12–14,32,33] to extract interpretable directions from latent space, manual annotations or outright labels must be added as supervision conditions. InterfaceGAN [13] is a classical supervised approach to semantic face editing by interpreting the latent semantics discovered by GANs. InterfaceGAN allows for exact control of facial characteristics (e.g., gender, age, expression, glasses). However, it also necessitates sampling a sizable amount of labeled data with the aid of an attribute predictor that has already been trained. In StyleFlow [32], labels are used in conjunction with a continuous normalized flow technique to localize the semantic directions in GANs. Additionally, methods [8,12,14,33] use pre-trained semantic predictors to identify interpretable semantics in the latent space. For instance, [12] finds semantic directions in latent space containing scene information by using target detectors to locate entity classes, attributes, and structural information. Even though supervised algorithms can discover interpretable directions of higher quality from latent space, they frequently necessitate the insertion of pricey external supervision conditions. Additionally, it can only find expected semantic features; it cannot find additional, unanticipated directions.

**Unsupervised Methods.** GANspace [22] performs Principal Component Analysis (PCA) on the latent space of pre-trained StyleGAN and BigGAN models without any constraint to find major interpretable directions. SeFa [23] discovers the latent semantic directions that GAN has learned by decomposing the model weights and examining the connection between image changes and internal representations. It does not depend on any training or labeling. In recent years, three approaches [34–36] capture key interpretable directions from a pre-trained GAN model. They all involved training a reconstructor and a direction matrix. Specifically, the reconstructor anticipates the changes received from the direction matrix to predict interpretable directions and displacements, whereas the direction matrix is utilized to detect semantic changes in latent space. They all use an unsupervised method to extract interpretable directions from the latent space of the GAN, but this also depends on how well the pre-trained GAN performed. Similarly, latentCLR [37] also uses training a direction matrix to discover interpretable directions from the pre-trained GAN model. However, the difference is that it employs a self-supervised contrast learning loss function to optimize training. These methods are post-processing techniques, meaning that they must get a pre-trained generative network. Only that can they find semantics, even though they do not necessitate the additional insertion of supervision conditions. As a result, their effectiveness heavily depends on how well the pre-trained GAN performs, and they also need to be operated in two steps. Instead of depending on a pre-trained GAN model, our approach only needs one operation step-capturing interpretable directions from changes in the samples during the GAN training.

## 2.3   Disentanglement Learning with Orthogonal Regularization

Many studies [38–40] have started with regularization to achieve disentanglement, aiming to enhance disentanglement, including regularization in the network training procedure. InfoGAN [38] enables variables to have interpretable information by constraining the relationship between latent code and generated results. Peebles et al. [39] proposes to include the regularization term Hessian Penalty in the generative model, encouraging a generative model's Hessian with regard to its input to be diagonal. So it can be used to find interpretable directions in BigGAN's latent space in an unsupervised fashion. In order to encourage the learnt representations to be orthogonal, D Wang et al. [41] implement orthogonal regularization on the weighting matrices of the model in the manner of $\left\| W^T W - I \right\|_2$, where W is the weight matrix and I is an identity matrix. Furthermore, Bansal et al. [42] introduced another form of regularization by considering both $\left\| W^T W - I \right\|_2$ and $\left\| W W^T - I \right\|_2$. However, the authors also note that this format does not always perform better than $\left\| W^T W - I \right\|_2$ and even performs worse on specific tasks, as evidenced by experimental findings.

EigenGAN [24] also uses regularization in the form of $\left\| W^T W - I \right\|_2$ in the subspace. According to the experimental findings, improving the disentanglement in the discovered interpretability direction is limited by adding the regularization of [41] alone. Thus, we introduce Orthogonal Jacobian regularization (OroJaR) [27], which limits the orthogonal characteristics of each input dimension in the model to disentangle the model by constraining the orthogonal properties of each dimension of the input between the changes induced by the output. In contrast to earlier approaches, the OroJaR can enable the generative model to learn disentangled variants more effectively.

## 3   Methods

### 3.1   Overview

Moving in latent space along a specific interpretable direction can get the visual effect after changing the corresponding semantic properties. Our goal is to discover some interpretable directions in GAN learning specific domain knowledge. Figure 2 depicts the overall framework of our model. Firstly, the latent code $z \in Z = \mathcal{N}(\mu, \sigma^2)$ is obtained by random sampling, where $z = [z_1, z_2, \ldots, z_l]$ and l is the number of generator layers. Following that, mapping network transformation is used to obtain the latent vector $w_i = f(z_i)$, where $w = [w_1, w_2, \ldots, w_l], w_i \in \mathcal{W} \subseteq \mathbb{R}^l$. Here, $f(\cdot)$ represents the mapping network

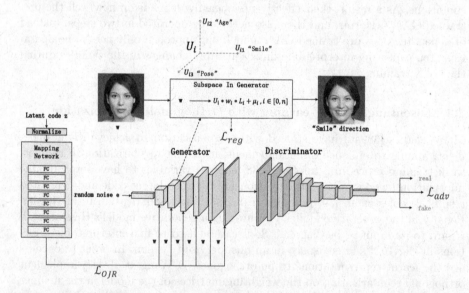

**Fig. 2.** The architecture of proposed method, which composed of a latent mapping network, one generator and discriminator. The randomly sampled latent code $z$ is converted into the vector $w$ by the latent mapping network and fed into the generator along with the randomly sampled noise. The subspace model of the generator learns interpretable directions from the sample variations during training.

implemented using the multilayer perceptron, and each $z_i$ shares the same f($\cdot$). As a result, we train just one latent mapping network. The generator G($\cdot$) receives the w vector as input to produce the fake image $x_{fake}$, which is then supplied to the discriminator D($\cdot$) with the genuine image $x_{real}$ for judgment. Eventually, the generator subspace learns interpretable directions from the changes in the sample distribution during training.

## 3.2    Layer-Wise Semantic Discovering Model

Numerous research [25,26] have shown that the space $W$ of styleGAN is rich in disentangled semantics. Moreover, the intermediate latent space can learn more semantically disentangled information than the original latent space. Therefore, inspired by [3], we convert the model input to the intermediate latent space. First, the latent code $z = [z_1, z_2, \ldots, z_l]$ is randomly sampled for each layer of the generator's subspace where the subscript $\ell$ denotes the number of generator layers. Furthermore, the $z$ is normalized and input to the latent mapping network. Additionally, the latent mapping network outputs a vector w $= [w_1, w_2, \ldots, w_l]$ of the space $W$ without altering the latent code's size. The w-vector is then individually input to each stage of the generator. This design choice is beneficial for the final disentangled interpretable directions. Moreover, our generative adversarial model, which draws inspiration from [3,12,24,43], adopts the layer-wise design concept. And the StyleGAN [3] and BigGAN [43]) also introduce it to improve the training stability and synthesis quality. The layer-wise GANs input constants at the first layer and latent code at each subsequent layer, in contrast to traditional GANs that only input latent code at the first layer. Like [24], we feed $w$ into each generator layer and random noise into the first layer. Using the FFHQ dataset as an example, the generator takes the random noise $\epsilon \subseteq \mathbb{R}^{512}$ in the first layer and $w_i$ in each layer as input and the synthetic image $x_{fake}$ as output. The discriminator facilitates the adversarial training by judging $x_{fake} = G([w_1, w_2, \ldots, w_l])$ and $x_{real}$, which eventually enables the generator to synthesize high-fidelity face images. The generator's subspace model learns the interpretable direction in the face images unsupervised from changes in the sample distribution during adversarial training. Notably, in contrast to the original latent code z, which obeys a Gaussian distribution, [25] states that the distribution of the intermediate latent code w cannot be explicitly modeled. Therefore, since the latent mapping network converts z into w, the changes learned by the subspace model will be more disentangled thanks to the distribution of w.

Similar to [24], we set up a subspace structure $M_i = [U_i, L_i, \mu_i], i \in [0, l]$ in each generative network layer to capture interpretable directions, where

- $U$ is the orthogonal basis of the subspace, which aims to discover interpretable directions in latent space. $U_i = [U_{i1}, U_{i2}, \ldots, U_{id}]$, where $d$ denotes the dimension of the subspace. Besides, it also represents the number of semantic directions discovered. Each basis vector $U_{ij} \in \mathbb{R}^{H_i \times W_i \times C_i}$ is used to discover one interpretable direction.

- $L$ is the importance matrix $L_i = \text{diag}\,(l_{i1}, l_{i2}, \ldots, l_{id})$, where the absolute value of $l_{ij}$ indicates the importance of $U_{ij}$ for the semantic change at level $i^{th}$.
- $\mu$ denotes the starting point of subspace operations.

The above three parameter values change with the training of the model, and when the training of the model is completed, we need to edit the interpretability direction found by $U_{ij}$. Firstly, the latent code $z = [z_1, z_2, \ldots, z_l]$ is randomly sampled and converted to vector $w = f(z)$. Then it is inputted into the subspace of each layer of the generator, activating $U_i$ in the particular $i^{th}$ layer and calculating to get the subspace coordinate points, as shown in Eq. (1).

$$\omega_i = U_i * w_i * L_i + \mu_i = \sum_{j}^{d} U_{ij} * w_{ij} * l_{ij} + \mu_i \tag{1}$$

Then $\omega_i$ is added to the network features in the $i^{th}$ layer for calculation, which determines the semantic changes in the $i^{th}$ layer of the generator.

### 3.3   Orthogonal Jacobian Regularization

In terms of disentanglement, supervised methods are typically more effective than unsupervised methods for finding interpretable directions in GAN. It is so that methods with additional supervised conditions can find directions that are more different from other directions to achieve disentanglement since they are better at identifying the target interpretable directions. Some methods increase the regularization of the training in order to achieve disentanglement [38, 39]. Additionally, EigenGAN [24] incorporates the Hessian penalty [39] in training. However, this is insufficient for discovering interpretable semantics in GAN since many semantics-induced changes are usually spatially dependent (e.g., Pose, Hairstyle, etc.). It is not sufficient to constrain each element alone. Instead, [27] proposed that constraining the changes induced by each latent factor in a holistic manner can achieve better disentanglement.

Inspired by this, we also introduce Orthogonal Jacobian regularization [27] in the model to achieve disentanglement. Let $h_i = G_i(w_i)$ in the $i^{th}$ layer of the generator, where $h_i$ is the network feature and $G_i$ denotes the output of the $i^{th}$ layer. For the $w_i = [w_{i1}, w_{i2}, \ldots, w_{id}]$ at layer $i^{th}$, to make $w_{ij}$ and $w_{iq}$, $j \neq q, j, q \in [0,\,d]$ the changes induced in the output of the $i^{th}$ layer independent, it is necessary to make the Jacobian vectors in each dimension of the input orthogonal to each other, as shown in Eq. (2).

$$\left[\frac{\partial G_i}{\partial w_{ij}}\right]^T \frac{\partial G_i}{\partial w_{iq}} = 0 \tag{2}$$

The orthogonality of the Jacobian vectors of $w_{ij}$ and $w_{iq}$ implies that they are also uncorrelated. Similar to [27], we consider using Orthogonal Jacobian regularized loss functions for all input dimensions to help the model learn to disentangle interpretable directions. As shown in Eq. (3).

$$L_{OJR} = \sum_{i=1}^{l} \sum_{j=1}^{d} \sum_{j \neq q} \left| \left[ \frac{\partial G_i}{\partial w_{ij}} \right]^T \frac{\partial G_i}{\partial w_{iq}} \right|^2 \tag{3}$$

where $l$ represents the generator's overall number of layers.

**Optimization Objective.** In addition, to further improve the model's effectiveness in finding the disentangled interpretable directions, we add the Hessian penalty [39] to the model for constraining the vectors $w$ of the input generators. Also, inspired by [41], an orthogonal loss function for $U$ is introduced in order to achieve the disentanglement between the interpretable directions found by $U$, as shown in Eq. (4),

$$\mathcal{L}_{re.g.} = \left\| U_i^T U_i - I \right\|^2 \tag{4}$$

Therefore, the loss functions of the generator and discriminator are shown in Eq. (5) and Eq. (6).

$$\mathcal{L}_G = \mathcal{L}_{adv\_G} + \mathcal{L}_{hes} + \mathcal{L}_{reg} + \mathcal{L}_{OJR} \tag{5}$$

$$\mathcal{L}_D = \mathcal{L}_{adv\_D} \tag{6}$$

where $\mathcal{L}_{adv\_G}$ and $\mathcal{L}_{adv\_D}$ are consistent with the adversarial objective function of GAN [5], and $\mathcal{L}_{hes}$ denotes the Hessian penalty [39].

## 4 Experiments

### 4.1 Experiment Settings

**Datasets.** To assess the efficacy of our approach, we used the FFHQ [3], Danbooru2019 Portraits [44], and LSUN-Church datasets [45]. They include the Danbooru2019 Portraits dataset, which has 30,2652 anime face photos, and the FFHQ dataset, which has 70,000 HD resolution face photographs. We aim to demonstrate the interpretable directions discovered from the dataset and evaluate their disentanglement. We also apply the method to LSUN-Church and present the interpretable directions we find in scene photographs and animal face images further to illustrate the resilience and efficacy of our proposed method.

**Implementation Details.** We perform all experiments using the Pytorch toolbox on a single NVIDIA GeForce RTX 1080Ti 11 GB. We reduce the image's resolution to 256 * 256 and increase the batch size to 8 due to the limitation of video memory size. And we select the Adam algorithm as the optimizer and set the initial learning rate to 1e-4.

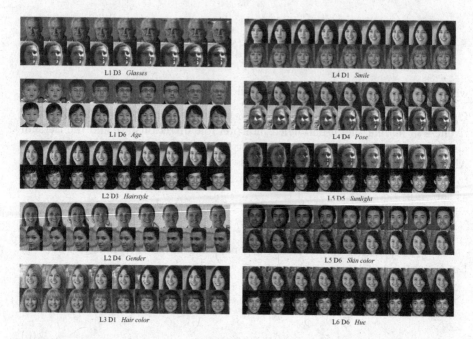

**Fig. 3.** Interpretable directions found in different layers of the FFHQ dataset [3]. The intensity of the attribute editing locates at $\in [-4\sigma, 4\sigma]$. And each dimension of the orthogonal basis corresponds to a specific semantic direction. We only show a few of the most meaningful attributes in the figure.

## 4.2　Non-trivial Visual Effects

First, we show the interpretable directions learned by our method for each layer of the subspace model during the GAN training. Figure 3 displays some illustrations of interpretable directions that the model discovered while learning about faces in the FFHQ dataset, where "Li Dj" denotes the $j^{\text{th}}$ dimension of the $i^{\text{th}}$ layer of the generator network. In addition, a larger value of $i$ indicates a shallower network layer. By setting $x_{\text{shift}} \in [-4\sigma, 4\sigma]$ and replacing the position of the dimension corresponding to that layer in the latent code z with the value of $x_{\text{shift}}$, we can initially activate a particular dimension of the subspace in a specific layer. Then the semantic editing result image is obtained by traversing the coordinate value in $[-4\sigma, 4\sigma]$ in a specific interpretable direction. As can be observed, progressing in the interpretable direction causes the image's general semantics to progressively shift in that way in order to provide an editing effect that is aesthetically acceptable to humans.

As shown in Fig. 3, the shallow subspaces of the model (layer5, layer6) tend to learn lower-level semantic attributes, such as L5D5 learning to "sunlight", the "skin tone" in L5D6, and the "hue" in L6D6. It is obvious that the shallow subspaces are more eager to discover color-related interpretable directions. The intermediate layer subspaces of the model are skewed to discover regional

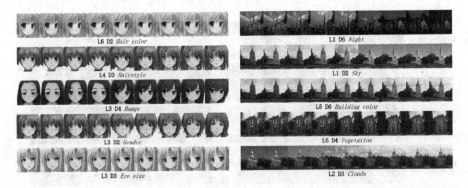

**Fig. 4.** Examples of interpretable directions found on the Anime face dataset [44] (left) and the LSUN-Church dataset [45] (right)

structural changes as the number of layers deepens. For instance, L3 learns the "hair color" attribute while L4 finds changes in position. We discover the deep subspace's propensity to discover high-level interpretable directions related to the abstract features of the deep network layer of the generator learning face knowledge. As seen in the figure, the deep subspace discovers the face's high-level semantic features, e.g., L0 and L1 found attributes such as "glasses", "gender", "hairstyle", and "age". However, the deeper subspaces do not have the same degree of disentanglement as the shallow ones. For instance, the attribute "beard" also appears in the figure when the attribute "gender" advances in the direction of "male". It is probable that the semantic qualities of "male" that the generator learned always include the attribute "beard". As a result, both attributes frequently show up in specific interpretable directions.

To sum up, shallow subspaces in generators often learn low-level features. In contrast, deeper subspaces can find more intricate and high-level interpretable directions, which is in line with the conclusion of Yang Ceyuan [12]. In addition, the conclusion reached by Bau David [6] in GAN models exploring hierarchical semantics, that various layers in a GAN model can find different levels of semantics, is validated by our method since we also use layer-wise ideas in the building of GAN models. However, there is still some entanglement in the interpretability directions of the deeper subspaces because the deeper layers of generators typically learn more abstract features.

On the other hand, we also apply our method to the anime face dataset and scene dataset LSUN-Church in order to confirm the method's applicability. As seen in Fig. 4, we can also find FFHQ-like interpretable directions in the anime face dataset, e.g., "Gender", "Hairstyle", "Bangs", etc. Similar to the findings of FFHQ, anime discovers high-level attributes like "Hair color" in the deep subspace and low-level interpretable directions like "Gender" in the shallow subspace. In addition, the Church dataset contains interpretable directions, e.g., "Vegetation", "Clouds", "Night", "Sky", and "Building color".

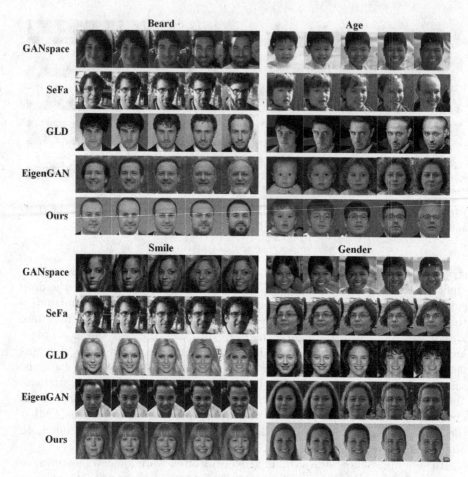

**Fig. 5.** Qualitative comparison among GANspace [22], SeFa [23], GLD [35], EigenGAN [24] and ours in four common interpretable directions

## 4.3  Comparison

In this section, to further demonstrate the effectiveness of our approach, we have chosen several classical unsupervised methods for qualitative and quantitative comparisons. Regarding the comparison method, we choose a few traditional unsupervised approaches, including GANspace [22], SeFa [23], GLD [35], and EigenGAN [24] for comparison.

### 4.3.1  Qualitative Analysis

We compare the interpretable directions found by different methods on the FFHQ face dataset. We download and immediately call the official models offered by GANspace, SeFa, GLD, and EigenGAN to prevent the effects of different machine performances on the model. Then, we call several methods to move the same step along that direction for the found interpretable direction to get the semantically edited results. Figure 5 lists a number of common semantic attributes on faces found in common by several methods. All methods achieve visual effects that are consistent with the target attributes. Compared with the other methods, EigenGAN and ours obtain smoother changes in visual effects in interpretable directions of movement. Since all methods use an unsupervised approach, the attribute disentanglement is not good enough. In terms of attribute disentanglement, however, our method is still superior to the others, e.g., the change of "Beard" is not entangled with "Gender" or "hair".

### 4.3.2 Quantitative Analysis

**FID Analysis.** For a quantitative comparison, we perform the following experiment. To evaluate the effect of the edited visuals obtained by moving along a specific interpretable direction on the image quality, we estimate the Fr'echet Inception Distance (FID) [46] for the images before and after editing. We chose an interpretable direction corresponding to the visual effect of the "smile" attribute for the calculation since all five methods can find it on FFHQ. Figure 6 illustrates the computation of the FID values between the original and edited images following several methods to change the semantics of the images along the semantic direction of the "smile" in different steps. The folded data in the figure demonstrates that our method achieves lower FID values at several edit intensities. Therefore, the edited image obtained by changing along the specific semantics direction found by our method is convincing in terms of visual effect.

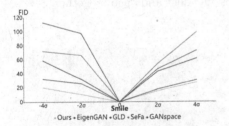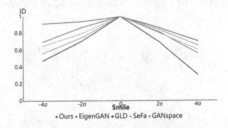

**Fig. 6.** FID plots for the semantic direction of "smile" produced by different methods. And a smaller FID value indicates that the editing has less impact on the image quality.

**Fig. 7.** ID plots for the semantic direction of "smile" produced by different methods. The closer the ID value is to 1, the better the identity information is preserved.

**ID Preservation Analysis.** On the other hand, we use ArcFace [47] to extract embedding vectors from edited images and compare identity preservation to the other four methods to assess our method's effect on the identity information of the original face after changing the semantics along the interpretable direction. This is a rather fair assessment of how well each method preserves the original facial identity information features while changing the semantics of the image.

Similar to the evaluation scheme of FID, we still chose the semantic direction of "smile". We calculate the identity similarity before and after changing the semantics of the face images in different steps along the "smile" direction (the value closer to 1 means the identity information is better preserved). Figure 7 illustrates that the ID value calculated by our method is closer to 1 than the other methods, thus proving the validity of preserving identity information.

**Re-scoring Analysis.** The disentanglement of the semantic attribute is also essential in the task of interpretable direction discovery. Therefore, the evaluation of several methods for semantic attribute disentanglement is also an indispensable part. We perform rescoring analysis with the help of a well-trained attribute predictor [23], which can recognize 40 facial attributes on the celebA [8] dataset. Specifically, an interpretable direction is first selected for shifting to change the image semantics. The edited image is then scored using the attribute predictor. Attribute disentanglement can be assessed numerically based on the changes in the scores of other semantic directions throughout the shift to a particular direction, in addition to qualitatively analyzing whether the identified direction accurately represents the relevant semantic attribute. Figure 8 shows the test results of EigenGAN (Fig. 8a) and our method (Fig. 8b). From the results in the figure, it is not difficult to conclude the observations: (i) Like EigenGAN, the interpretable directions discovered by our method do control the change of specific semantics. (ii) Some attributes are more easily entangled, such as "gender" and "beard", which are associated with the performance of both methods, probably since "beard" is often the same as "male" when GAN learns facial features. (iii) Our method performs better than EigenGAN in terms of attribute disentanglement, e.g., EigenGAN entangles "gender" and "age" together.

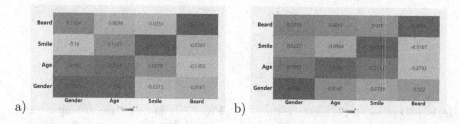

**Fig. 8.** The results of re-scoring analysis after training EigenGAN [24] (a) and our method(b) on the FFHQ dataset [3]. The data in the table shows how the scores of other semantic features change after moving in a specific semantic direction.

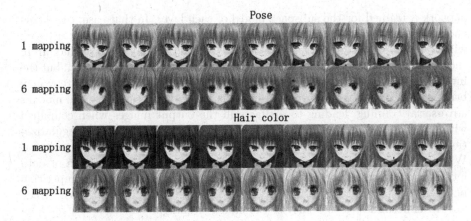

**Fig. 9.** The visual effects of two strategies for editing the "Pose" and "Hair color" attributes with training one mapping network and six mapping networks.

## 4.4   Ablation Studies

In our method design, the latent code z of each layer is obtained by random sampling and then input to the mapping network to transform to the intermediate latent space to get the w vector. In our architecture, only one mapping network is trained, and the latent code z of each layer in GAN is transformed through this mapping network. However, since each subspace in the GAN learns different levels of feature for each layer, we propose the following conjecture. Suppose a separate mapping network is trained for the subspace of each layer. Is there an improvement in the editing effect for the interpretable directions learned in the subspace of each layer? In this section, we design an ablation experiment to verify whether the above conjecture is feasible.

First, we compare two strategies: training six mapping networks (our method is designed with six layers of subspaces) versus training just one mapping network. Finally, we compare the editing effects in the interpretability direction for both strategies discovered in the same layer and dimension.

Figure 9 shows the visual effect of editing the attribute "pose" found in L4D1 and the attribute "hair color" found in L5D4 for the above two strategies. Although both methods identify the same interpretable direction in the same dimension at the same layer and the disentanglement and semantic effects are likewise positive, it is clear that the visual effect produced by the training procedures for six mapping networks contains glaring artifacts.

It is well known that the adversarial training process of generative adversarial networks is prone to instability for various reasons compared to other models. Inspired by [48,49], the following two conjectures were obtained regarding the reasons for the causes of artifacts when training six mapping networks. (i) Networks with more layers are typically stronger than networks with fewer tiers in the neural network. In contrast, our model has a simpler structure and fewer network layers (the generator has only six layers). Suppose the additional mapping network is trained for the subspace model of each layer. In that case, the difference in the distribution of w-vectors obtained between each mapping network will become increasingly large. Moreover, J Wulff et al. [50] also conclude from their analysis that the original noise space obeys a Gaussian distribution, but the intermediate latent space W cannot be clearly modeled. The above strategy may bring complex negative information into the model and destabilize the model's adversarial training, leading to artifacts in the output images when combined with the layer-wise design idea of our method. (ii) We add Jacobian regularization to the model to impose constraints on the input of the entire generator. When combined with the analysis in the first point, the introduction of data into the model with large distributional differences may result in competition between the Jacobian regularization loss and the adversarial training loss, which in turn undermines the stability of the model training.

## 5   Conclusion

This paper suggests a method for finding disentangled interpretable directions in GAN's latent space in an unsupervised form. We adopt the layer-wise idea to construct the GAN and add the subspace model to the generator to capture the interpretable directions. Since space $W$ is rich in disentangled semantics, we also introduce a latent mapping network to convert the model input to the intermediate latent vector $w$. In addition, to further achieve well disentanglement, we add the Orthogonal Jacobian regularization to the model to impose constraints on the overall model input. According to the experimental results, compared with existing methods, ours achieves excellent improvement in the disentanglement effect, both in terms of qualitative analysis of the editing effect in the interpretable direction and quantitative analysis of the degree of disentanglement.

**Acknowledgments.** This work was supported in part by the National Natural Science Foundation of China under grant 62072169, and Natural Science Foundation of Hunan Province under grant 2021JJ30138.

## References

1. Miyato, T., Kataoka, T., Koyama, M., Yoshida, Y.: Spectral normalization for generative adversarial networks. arXiv preprint arXiv:1802.05957 (2018)
2. Karras, T., Aila, T., Laine, S., Lehtinen, J.: Progressive growing of GANs for improved quality, stability, and variation. arXiv preprint arXiv:1710.10196 (2017)
3. Karras, T., Laine, S., Aila, T.: A style-based generator architecture for generative adversarial networks. In: Proceedings of the IEEE/CVF Conference on Computer Vision and Pattern Recognition, pp. 4401–4410 (2019)
4. Karras, T., Laine, S., Aittala, M., Hellsten, J., Lehtinen, J., Aila, T.: Analyzing and improving the image quality of styleGAN. In: Proceedings of the IEEE/CVF Conference on Computer Vision and Pattern Recognition, pp. 8110–8119 (2020)
5. Goodfellow, I., et al.: Generative adversarial nets. In: Advances in Neural Information Processing Systems, vol. 27 (2014)

6.  Bau, D., Zhu, J.-Y., Strobelt, H., Lapedriza, A., Zhou, B., Torralba, A.: Understanding the role of individual units in a deep neural network. Proc. Natl. Acad. Sci. **117**(48), 30071–30078 (2020)
7.  Zhou, B., Bau, D., Oliva, A., Torralba, A.: Interpreting deep visual representations via network dissection. IEEE Trans. Pattern Anal. Mach. Intell. **41**(9), 2131–2145 (2018)
8.  Bau, D., et al.: GAN dissection: visualizing and understanding generative adversarial networks. arXiv preprint arXiv:1811.10597 (2018)
9.  Zeiler, M.D., Fergus, R.: Visualizing and understanding convolutional networks. In: Fleet, D., Pajdla, T., Schiele, B., Tuytelaars, T. (eds.) ECCV 2014. LNCS, vol. 8689, pp. 818–833. Springer, Cham (2014). https://doi.org/10.1007/978-3-319-10590-1_53
10. Zhou, B., Khosla, A., Lapedriza, A., Oliva, A., Torralba, A.: Object detectors emerge in deep scene CNNs. arXiv preprint arXiv:1412.6856 (2014)
11. Bau, D., Zhou, B., Khosla, A., Oliva, A., Torralba, A.: Network dissection: quantifying interpretability of deep visual representations. In: Proceedings of the IEEE Conference on Computer Vision and Pattern Recognition, pp. 6541–6549 (2017)
12. Yang, C., Shen, Y., Zhou, B.: Semantic hierarchy emerges in deep generative representations for scene synthesis. Int. J. Comput. Vision **129**(5), 1451–1466 (2021). https://doi.org/10.1007/s11263-020-01429-5
13. Shen, Y., Gu, J., Tang, X., Zhou, B.: Interpreting the latent space of GANs for semantic face editing. In: Proceedings of the IEEE/CVF Conference on Computer Vision and Pattern Recognition, pp. 9243–9252 (2020)
14. Goetschalckx, L., Andonian, A., Oliva, A., Isola, P.: GANalyze: toward visual definitions of cognitive image properties. In: Proceedings of the IEEE/CVF International Conference on Computer Vision, pp. 5744–5753 (2019)
15. Jiang, Y., Huang, Z., Pan, X., Loy, C.C., Liu, Z.: Talk-to-edit: fine-grained facial editing via dialog. In: Proceedings of the IEEE/CVF International Conference on Computer Vision, pp. 13799–13808 (2021)
16. Patashnik, O., Wu, Z., Shechtman, E., Cohen-Or, D., Lischinski, D.: StyleCLIP: text-driven manipulation of styleGAN imagery. In: Proceedings of the IEEE/CVF International Conference on Computer Vision, pp. 2085–2094 (2021)
17. Couairon, G., Grechka, A., Verbeek, J., Schwenk, H., Cord, M.: FlexIT: towards flexible semantic image translation. In: Proceedings of the IEEE/CVF Conference on Computer Vision and Pattern Recognition, pp. 18270–18279 (2022)
18. Radford, A., et al.: Learning transferable visual models from natural language supervision. In: International Conference on Machine Learning, pp. 8748–8763. PMLR (2021)
19. Yao, X., Newson, A., Gousseau, Y., Hellier, P.: A latent transformer for disentangled face editing in images and videos. In: Proceedings of the IEEE/CVF International Conference on Computer Vision, pp. 13789–13798 (2021)
20. Shoshan, A., Bhonker, N., Kviatkovsky, I., Medioni, G.: GAN-control: explicitly controllable GANs. In: Proceedings of the IEEE/CVF International Conference on Computer Vision, pp. 14083–14093 (2021)
21. Lang, O., et al.: Explaining in style: training a GAN to explain a classifier in StyleSpace. In: Proceedings of the IEEE/CVF International Conference on Computer Vision, pp. 693–702 (2021)
22. Härkönen, E., Hertzmann, A., Lehtinen, J., Paris, S.: GANspace: discovering interpretable GAN controls. Adv. Neural. Inf. Process. Syst. **33**, 9841–9850 (2020)

23. Shen, Y., Zhou, B.: Closed-form factorization of latent semantics in GANs. In: Proceedings of the IEEE/CVF Conference on Computer Vision and Pattern Recognition, pp. 1532–1540 (2021)
24. He, Z., Kan, M., Shan, S.: EigenGAN: layer-wise eigen-learning for GANs. In: Proceedings of the IEEE/CVF International Conference on Computer Vision, pp. 14408–14417 (2021)
25. Tov, O., Alaluf, Y., Nitzan, Y., Patashnik, O., Cohen-Or, D.: Designing an encoder for styleGAN image manipulation. ACM Trans. Graph. (TOG) **40**(4), 1–14 (2021)
26. Wu, Z., Lischinski, D., Shechtman, E.: StyleSpace analysis: disentangled controls for styleGAN image generation. In: Proceedings of the IEEE/CVF Conference on Computer Vision and Pattern Recognition, pp. 12863–12872 (2021)
27. Wei, Y., et al: Orthogonal jacobian regularization for unsupervised disentanglement in image generation. In: Proceedings of the IEEE/CVF International Conference on Computer Vision, pp. 6721–6730 (2021)
28. Roth, K., Lucchi, A., Nowozin, S., Hofmann, T.: Stabilizing training of generative adversarial networks through regularization. In: Advances in Neural Information Processing Systems, vol. 30 (2017)
29. Mescheder, L., Geiger, A., Nowozin, S.: Which training methods for GANs do actually converge? In: International Conference on Machine Learning, pp. 3481–3490. PMLR (2018)
30. Nowozin, S., Cseke, B., Tomioka, R.: f-GAN: training generative neural samplers using variational divergence minimization. In: Advances in Neural Information Processing Systems, vol. 29 (2016)
31. Mao, X., Li, Q., Xie, H., Lau, R.Y., Wang, Z., Paul Smolley, S.: Least squares generative adversarial networks. In: Proceedings of the IEEE International Conference on Computer Vision, pp. 2794–2802 (2017)
32. Abdal, R., Zhu, P., Mitra, N.J., Wonka, P.: StyleFlow: attribute-conditioned exploration of styleGAN-generated images using conditional continuous normalizing flows. ACM Trans. Graph. (TOG) **40**(3), 1–21 (2021)
33. Shen, Y., Yang, C., Tang, X., Zhou, B.: InterFaceGAN: interpreting the disentangled face representation learned by GANs. IEEE Trans. Pattern Anal. Mach. Intell. **44**, 2004–2018 (2020)
34. Cherepkov, A., Voynov, A., Babenko, A.: Navigating the GAN parameter space for semantic image editing. In: Proceedings of the IEEE/CVF Conference on Computer Vision and Pattern Recognition, pp. 3671–3680 (2021)
35. Voynov, A., Babenko, A.: Unsupervised discovery of interpretable directions in the GAN latent space. In: International Conference on Machine Learning, pp. 9786–9796. PMLR (2020)
36. Tzelepis, C., Tzimiropoulos, G., Patras, I.: WarpedGANSpace: finding nonlinear RBF paths in GAN latent space. In: Proceedings of the IEEE/CVF International Conference on Computer Vision, pp. 6393–6402 (2021)
37. Yüksel, O.K., Simsar, E., Er, E.G., Yanardag, P.: LatentCLR: a contrastive learning approach for unsupervised discovery of interpretable directions. In: Proceedings of the IEEE/CVF International Conference on Computer Vision, pp. 14263–14272 (2021)
38. Chen, X., Duan, Y., Houthooft, R., Schulman, J., Sutskever, I., Abbeel, P.: InfoGAN: interpretable representation learning by information maximizing generative adversarial nets. In: Advances in Neural Information Processing Systems, vol. 29 (2016)

39. Peebles, W., Peebles, J., Zhu, J.-Y., Efros, A., Torralba, A.: The hessian penalty: a weak prior for unsupervised disentanglement. In: Vedaldi, A., Bischof, H., Brox, T., Frahm, J.-M. (eds.) ECCV 2020. LNCS, vol. 12351, pp. 581–597. Springer, Cham (2020). https://doi.org/10.1007/978-3-030-58539-6_35

40. Zhu, X., Xu, C., Tao, D.: Learning disentangled representations with latent variation predictability. In: Vedaldi, A., Bischof, H., Brox, T., Frahm, J.-M. (eds.) ECCV 2020. LNCS, vol. 12355, pp. 684–700. Springer, Cham (2020). https://doi.org/10.1007/978-3-030-58607-2_40

41. Wang, D., Cui, P., Ou, M., Zhu, W.: Deep multimodal hashing with orthogonal regularization. In: Twenty-Fourth International Joint Conference on Artificial Intelligence (2015)

42. Bansal, N., Chen, X., Wang, Z.: Can we gain more from orthogonality regularizations in training deep networks? In: Advances in Neural Information Processing Systems, vol. 31 (2018)

43. Brock, A., Donahue, J., Simonyan, K.: Large scale GAN training for high fidelity natural image synthesis. arXiv preprint arXiv:1809.11096 (2018)

44. Branwen, G., Gokaslan, A.: Danbooru 2019: a large-scale crowdsourced and tagged anime illustration dataset (2019)

45. Yu, F., Seff, A., Zhang, Y., Song, S., Funkhouser, T., Xiao, J.: LSUN: construction of a large-scale image dataset using deep learning with humans in the loop. arXiv preprint arXiv:1506.03365 (2015)

46. Heusel, M., Ramsauer, H., Unterthiner, T., Nessler, B., Hochreiter, S.: GANs trained by a two time-scale update rule converge to a local nash equilibrium. In: Advances in Neural Information Processing Systems, vol. 30 (2017)

47. Deng, J., Guo, J., Xue, N., Zafeiriou, S.: ArcFace: additive angular margin loss for deep face recognition. In: Proceedings of the IEEE/CVF Conference on Computer Vision and Pattern Recognition, pp. 4690–4699 (2019)

48. Liang, J., Zeng, H., Zhang, L.: Details or artifacts: a locally discriminative learning approach to realistic image super-resolution. In: Proceedings of the IEEE/CVF Conference on Computer Vision and Pattern Recognition, pp. 5657–5666 (2022)

49. Ledig, C., et al.: Photo-realistic single image super-resolution using a generative adversarial network. In: Proceedings of the IEEE Conference on Computer Vision and Pattern Recognition, pp. 4681–4690 (2017)

50. Wulff, J., Torralba, A.: Improving inversion and generation diversity in StyleGAN using a gaussianized latent space. arXiv preprint arXiv:2009.06529 (2020)

# ASNN: Accelerated Searching for Natural Neighbors

Dongdong Cheng[1,2], Jiangmei Luo[3](✉), Jinlong Huang[1], and Sulan Zhang[1]

[1] College of Big Data and Intelligent Engineering, Yangtze Normal University, Chongqing, China
{cdd,h.jinlong}@yznu.edu.cn, slzhang@cqu.edu.cn
[2] Chongqing Key Laboratory of Computational Intelligence, Chongqing University of Posts and Telecommunications, Chongqing, China
[3] College of Computer Science, Chongqing University, Chongqing, China
jmluo@cqu.edu.cn

**Abstract.** How to set $k$ value for k-nearest neighbors is a primary problem in machine learning, pattern recognition and knowledge discovery. Natural neighbor (NaN) is an adaptive neighbor concept for solving this problem, which combines k-nearest neighbors and reverse k-nearest neighbors to adaptively obtain $k$ value. It has been proven effective in clustering analysis, classification and outlier detection. However, the existing algorithms for searching NaN all use a global search strategy, which increases unnecessary consumption of time on non-critical points. In this paper, we propose a novel accelerated algorithm for searching natural neighbor, called ASNN. It is based on the fact that if the remote objects have NaNs, others certainly have the NaNs. The main idea of ASNN is that it first extracts remote points, then only searches the neighbors of remote points, instead of all points, so that ASNN can quickly obtain the natural neighbor eigenvalue (NaNE). To identify the remote objects, ASNN only searches the 1-nearest neighbor for each object with kd-tree, so its time complexity is reduced to $O(nlogn)$ from $O(n^2)$, and the local search strategy makes it run faster than the existing algorithms. To illustrate the efficiency of ASNN, we compare it with three existing algorithms NaNs, kd-NaN and FSNN. The experiments on synthetic and real datasets tell that ASNN runs much faster than NaNs, kd-NaN and FSNN, especially for datasets with large scale.

**Keywords:** Natural neighbor · Remote point · Remote region neighbor · Local stable state

## 1 Introduction

$K$-nearest neighbors and $\varepsilon$-neighborhood [18] are two representative neighbor concept, which are used to model the data and have been widely used in machine learning, pattern recognition, and knowledge discovery. By employing them, we can discover the potential relationships between objects. The $k$-nearest neighbors

T. Li et al. (Eds.): BigData 2022, CCIS 1709, pp. 40–56, 2022.
https://doi.org/10.1007/978-981-19-8331-3_3

(KNN) of an object is defined as the $k$ nearest objects closest to this object. Based on the concept, KNN classifier has become one of the simplest and most common classifier [1]. In the density-based clustering algorithms [9,16], researchers defined $\varepsilon$-neighborhood to calculate the density of objects. It is defined as a set of objects in the radius $\varepsilon$ of the object. Chen et al. [3,4] have proven that $\varepsilon$-neighborhood problem can be transformed into the k-nearest neighbors problem. However, $k$-nearest neighbors hides an unavoidable limitation: the choice of the parameter $k$.

Determining the optimum neighbor's parameter $k$ is one of the most studied problems in depth since the nearest neighbor method was proposed. The dilemma of neighborhood selection can be summarized in two aspects: the uncertainty of the dataset, and the sensitivity of neighbor's parameters. The characteristics of data distribution are usually completely unknown before the data are analyzed. Furthermore, the selection of neighbor's parameters also extremely depends on the knowledge and experiences of researchers. If the neighbor's parameter is too small, the number of neighbors of the data object will be less than the number of neighbors it should have, which causes the neighbor relationship to be ignored. In another opposite case, if the neighbor's parameter is too large, the boundaries between the classes will be blurred, and mislead the decision. Therefore, the parameter of $k$ is usually estimated by repeated experiments.

Natural neighbor (NaN) was creatively proposed [25] against the problem of neighbor's parameter selection. The natural neighbor eigenvalue (NaNE) of the natural neighbor has high-quality reference significance for determining the optimum parameter. In addition, this idea completely gets rid of the shortcomings of parameter dependence, and at the same time, the natural neighbors of each data point can be found spontaneously. It is widely used in clustering, classification and outlier detection. In clusteirng analysis, Cheng et al. [5–8] utilize NaN to define local density peaks or dense cores, which are used to reduce the scale of data. Zhang et al. [23] introduce NaN to construct a sparse graph and use it to cluster data. In classification, Li et al. [13,14] propose an error detection method based on NaN to detect noisy and borderline examples in imbalanced classification. Yang et al. [21] use natural neighborhood graph to eliminate noisy and redundant objects in training set. In outlier detection, Huang et al. propose natural outlier factor (NOF) [11] and ROCF [10] based on NaN to detect outliers or outlier clusters without setting parameters. In [19], the authors use natural neighbor concept to adaptively obtain natural value and propose a natural neighbor-based outlier detection algorithm. Besides, it is also used in oversampling [17], 3D point cloud data simplification [12,20], semi-supervised [24], and so on.

However, the earlier algorithms for searching natural neighbors [25] require continuously expanding the neighbor search round $r$ until the natural stable state is reached. At each iteration, KNN and RKNN of all data points are computed. kd-NaN [8] introduces kd-tree to search k-nearest neighbors, instead of the distance matrix. Therefore, the time complexity of kd-NaN is $O(nlogn)$. FSNN [22] focuses on determining NaNE quickly to speed up the search of natural neighbors. It first computes the distance matrix and obtains the most remote neighbor for each object according to the distance matrix. Then, the most remote neighbor with the most times in other objects becomes the most the remote point. Finally, the maximum rank of the most remote point is the NaNE. Obviously, the earlier algorithms for searching natural neighbors are global search algorithms. It increases unnecessary consumption of time in traversing many non-critical data objects, especially on large datasets.

To overcome this issue, a novel accelerated algorithm for searching natural neighbor (ASNN) is proposed in this paper. ASNN is based on the fact that if remote points have natural neighbors, others certainly have natural neighbors. Firstly, ASNN extracts the remote points by the defined remote radius. Secondly, neighbors of each remote point are searched in the defined remote neighborhood until the local stable state is reached. The global stable state is formed when all remote points reach the local stable state. Finally, NaNE can be quickly obtained by finding the maximum search round. ASNN employs a local search strategy that just searches neighbors for remote points instead of all points to quickly obtain NaNE. In this way, the number of query points is reduced drastically, the search area is narrowed and the time consumption is saved. The main contributions of this paper are summarized as follows:

(1) An accelerated algorithm for searching natural neighbor is proposed. It employs a local search strategy to obtain NaNE and natural neighbors.
(2) We define remote points and employ a fast method to find them. The method is based on remote radius, remote region neighbor and remote neighborhood. The proposed remote point can be automatically adapted to the distribution of a dataset without human intervention.
(3) The experiments on synthetic and real datasets show that the proposed algorithm runs much faster than the compared algorithms.

The remaining content of this paper is organized as follows. Section 2 reviews the related work about k-nearest neighbor, Reverse k-nearest neighbor and Natural neighbor. Section 3 introduces the proposed method in detail and Sect. 4 shows the experimental analysis. Finally, we conclude this work in Sect. 5.

## 2    Related Work

### 2.1    K-Nearest Neighbor

$K$-nearest neighbor (KNN for short) is a fundamental issue in various study fields. In 1951, Stevens [18] introduced an efficient algorithm for the problem by forming a subset of a point and its nearest neighbors. Given a set of $n$ data points $X = \{x_1, x_2, \cdots, x_n\}$ in $\mathbb{R}$, if $d(x_i, x_j)$ is the similarity (i.e., distance) between data object $x_i$ and $x_j$. The definition of $k$-nearest neighbor is as follows.

**Definition 1** ($k$-Nearest Neighbors). Given a query object $x_i$, the $k$-nearest neighbor search is to find point $x_j$ in $X$ with $d(x_i, x_j) \leq d(x_i, x_k)$, i.e., $NN_k(x_i) = \{x_j \in X \mid d(x_i, x_j) \leq d(x_i, x_k)\}$. $x_k$ has the $k$-th smallest distance between $x_i$ and $x_k$.

## 2.2    Reverse k-Nearest Neighbor

Reverse $k$-nearest neighbor (RKNN) [15] is a concept of the neighbor that has received widespread attention and application. Its history begins in 2000 with the fundamental work of Korn and Muthukrishnan. Following [15], the definition of reverse $k$-nearest neighbors is as follows.

**Definition 2** (Reverse $k$-nearest neighbors). Given a query object $x_i$, the reverse $k$-nearest neighbor search is to find a set of points that are regarded as one of the $k$-nearest neighbors of data object $x_i$, i.e., $RNN_k(x_i) = \{x_j \in X \mid \in x_i \in NN_k(x_j)\}$.

However, the parameter sensitivity of KNN and RKNN leads to the fact that determining the optimum value of the neighbor's parameter is a challenge.

## 2.3    Natural Neighbor

Zou and Zhu [26] firstly conduct the fundamental research in 2011, and then the theory of natural neighbor (NaN) is proposed in [25]. In [25], the authors also introduce $kd$-tree to reduces the time complexity of the original algorithm [8]. The main characteristics of the natural neighbors are that: (a) it is parameter-free; (b) it is more suitable for manifold data; (c) each data object has different numbers of natural neighbors; (d) the natural neighbor eigenvalue (NaNE) is a better choice of the parameter in KNN and RKNN; (e) it can detect outliers.

The thought of NaN is similar to the friendship of human society. Clearly, true friendships should be mutual, and if two people consider each other as friends, then they are true friends. When everyone in a community has a true friend, that community reaches a state of harmony. Similarly, in data space, if both point $x_i$ and point $x_j$ regarded each other as their own neighbors, then $x_i$ and $x_j$ are true friends (i.e., true neighbors). If every data object has a true friend, then the dataset reaches a harmonious state (i.e., the natural stable state in [25]), and the neighbor relationship at that time is the natural neighbor. The formal definition of the natural neighbor can be given as follows.

**Definition 3** (Natural neighbor). Given a set of $n$ data points $X = \{x_1, x_2, \cdots, x_n\}$ in $\mathbb{R}$ and $x_i, x_j \in X$. If $x_i$ belongs to the $supk$-th nearest neighbors of $x_j$ and $x_j$ belongs to the $supk$-th nearest neighbors of point $x_i$, then $x_i$ and $x_j$ are considered to be the natural neighbor of each other.

$$x_j \in NaN(x_i) \Leftrightarrow (x_i \in NN_{supk}(x_j)) \wedge (x_j \in NN_{supk}(x_i)) \tag{1}$$

where $NN_{supk}(x_i)$ is the set of $supk$ nearest neighbors of $x_i$. The objective of natural neighbor searching is to achieve a natural stable state, where each data object has at least one true friend (i.e., the mutual neighbor).

Algorithm 1 describes the process of natural neighbor searching. First, it calculates the Euclidean distance matrix. Then, it iteratively searches $r$ nearest neighbors and reverse $r$ nearest neighbors, counts the number of reverse $r$ nearest neighbors, and determine whether the searching state is stable. Before the searching state is stable, the natural neighbor of each data point is searched by continuously expanding the neighbor searching round $r$. Since NaNs need to calculate the distance matrix, its time complexity is $O(n^2)$.

The neighbor searching round $r$ at the stable state is called the natural neighbor eigenvalue (NaNE) $supk$, and it is defined as follows.

**Definition 4** (Natural neighbor eigenvalue $supk$). Natural neighbor eigenvalue is equal to the search round $r$ at the end of the search.

$$supk = r_{r \in n} \{r| \ (\forall x_i)(\exists x_j)(r \in n) \wedge (x_i \neq x_j)$$
$$\rightarrow (x_i \in NN_r(x_j) \wedge x_j \in NN_r(x_i) )\} \tag{2}$$

The existing works on natural neighbor tend to apply natural neighbors to different fields. Few of them improve the efficiency of natural neighbor searching. kd-NaN [8] replaces step 3 in Algorithm 1, that is, it search KNN and RKNN with kd-tree, instead of the distance matrix. Therefore, the time complexity of kd-NaN is $O(nlogn)$. FSNN [22] focuses on determining NaNE quickly to speed up the search of natural neighbors. It first computes the distance matrix and obtains the most remote neighbor for each object according to the distance matrix. The most remote neighbor with the most times in other objects becomes the most the remote point. Finally, the maximum rank of the most remote point is the NaNE. It directly finds NaNE through the distance matrix and avoids multiple searching KNN and RKNN for all points. Thus, it reduces the time for

---

**Algorithm 1.** Search for Natural neighbors (NaNS)

---

**Input:** Dataset $X = \{x_1, x_2, \cdots, x_n\} \in \mathbb{R}$.
**Output:** Natural neighbor $\boldsymbol{NaN}$, Natural Eigenvalue $supk$, the number of reverse neighbors $\boldsymbol{nb}$.
1: Initializing: $r \leftarrow 1$, $NN(x_i) \leftarrow \emptyset$, $RNN(x_i) \leftarrow \emptyset$, $\boldsymbol{NaN}(x_i) \leftarrow \emptyset$.
2: $n \leftarrow$ the total number of data points in $X$;
3: $\boldsymbol{dist} \leftarrow$ calculate the Euclidean distance matrix between each data point;
4: **while** true **do**
5:    **for** each data object $x_i \in X$ **do**
6:       Find $NN_r(x_i)$ and $RNN_r(x_i)$ by $\boldsymbol{dist}$;
7:       **if** $RNN_r(x_i) \neq \emptyset$ **then**
8:          $nb(x_i) \leftarrow$ the number of $RNN_r(x_i)$;
9:          $NaN(x_i) \leftarrow NN(x_i) \cap RNN(x_i)$;
10:      **end if**
11:    **end for**
12:    Compute $Numb$, the number of data point $x_i$ with $nb(x_i) = 0$;
13:    **if** $Numb == 0$ or $(Numb(r) == Numb(r-1)$ and $(r \geq 2))$ **then**
14:      **break**;
15:    **end if**
16:    $r \leftarrow r + 1$;
17: **end while**
18: $supk \leftarrow r$;
19: **return** $\boldsymbol{NaN}, \boldsymbol{nb}, supk$.

---

natural neighbor search. However, since it also requires computing the distance matrix, its time complexity is $O(n^2)$.

## 3    Proposed Method

### 3.1    Motivation and Theory

Natural neighbor [25] is a parameter-free and effective method, also can automatically determine the neighbor relationship between data objects, but iteratively and globally searching for neighbors for each data object is time-consuming, especially for large datasets. To address this issue, we propose an accelerated algorithm for searching natural neighbor, called ASNN.

Our method is inspired by sociology: if the person who is the hardest to find friends can find close friends, then it is easy for other individuals to get friends. Thinking about the same case in the data space, each data object is considered as an independent individual. Data objects in sparse areas have fewer neighbors with simple neighbor's relationships, while data objects in dense areas have more neighbors. Data objects in sparse areas are more difficult to find true friends (i.e., mutual neighbors) than those in dense areas. Clearly, consistent with the sociological theory above, when the data objects that are most difficult to have true friends have neighbors, other data objects that are easy to obtain true friends must have neighbors. The objective of natural neighbor searching is to find at least one mutual neighbor for each data object. Consequently, by focusing on data objects in sparse regions and finding at least one mutual neighbor for them, the natural neighbor eigenvalue can be determined more quickly.

In addition, to further narrow the search range and improve the search speed, in our method, the natural neighbor searching is changed from global search (i.e., search for all data objects) to local search (i.e., searching for key data objects). In ASNN, finding neighbors is a local problem. The neighbors of the data object will only appear in its vicinity and will not be located far away. Taking the dataset Gaussians as an example. The black dots, red pentagram $x_i$ and yellow pentagram $x_2$ in Fig. 1 are all independent data objects. There are two clusters, $C1$ and $C2$. For example, for $x_1$, the previous algorithm (Algorithm 1) calculates the distance between $x_1$ and all other data objects, leading to redundant and time-consuming in non-critical points. It can be observed from Fig. 1 that $x_1$'s neighbors are inevitably located in the vicinity, with a high probability in $C1$. It is almost impossible for the point in $C2$ (e.g., $x_2$) to be the nearest neighbor of $x_1$. Similarly, only data objects in $C2$ are potential neighbors of $x_2$. It is almost impossible for data objects in cluster $C1$ (e.g., $x_1$) to be nearest neighbors of $x_2$. Hence, the global neighbor search (finding neighbors for all data objects) is unnecessary. The proposed ASNN algorithm searches for neighbors only in the local area (i.e., the defined remote neighborhood), thus reducing the time overhead of natural neighbor search.

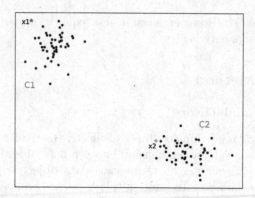

**Fig. 1.** Analysis of the ASNN. (Color figure online)

## 3.2 Accelerated Search of Natural Neighbors Algorithm

Given a set of data objects $X = \{x_1, x_2, \cdots, x_n\} \in \mathbb{R}$. For two object $x_i, x_j \in X$, $d(x_i, x_j)$ is the similarity between them. Unless otherwise specified, Euclidean distance is used in this paper. Some definitions of ASNN can be described as follows:

**Definition 5** (Remote radius). If data point $x_i$ has the nearest neighbor $NN_1(x_i)$ and the distance between them is $d(x_i, NN_1(x_i))$, then the remote radius is the maximum value in the distance list, abbreviated as $rd$. The equation of the remote radius is as follows:

$$rd = max\left(d\left(x_i, NN_1\left(x_i\right)\right)\right), i = 1, 2 \cdots, n \tag{3}$$

**Definition 6** (Remote neighborhood). Let object $x_i$ be the center and $rd$ be the radius, a closed circular region is formed. The above area is defined as the remote neighborhood of the data object $x_i$, and is denoted as $S(x_i)$.

**Definition 7** (Remote region neighbor). The set of data objects in the range of $S(x_i)$ is called the remote region neighbors $SN(x_i)$ of $x_i$. $|SN(x_i)|$ represents the number of remote region neighbors of $x_i$.

**Definition 8** (Remote point). If $|SN(x_i)|$ is smaller than the average number of the remote region neighbors of all data, then $x_i$ is regarded as a remote point, the opposite is a non-remote point. The formalized definition is:

$$R_i = \begin{cases} 1, |SN(x_i)| < \overline{SN} \\ 0, |SN(x_i)| \geq \overline{SN} \end{cases} \tag{4}$$

where $R_i$ is the identification of a remote point. If $R_i$ is 1, it means that $x_i$ is a remote point. If it is 0, it means that $x_i$ is a non-remote point. $\overline{SN}$ is the average number of remote region neighbors in the data set, and it is calculated as follows:

$$\overline{SN} = \frac{1}{n} \sum_{j=1}^{n} |SN(x_j)| \tag{5}$$

**Definition 9** (Local stable state). If a remote point $x_i$ reaches the local stable state, $\exists x_j \in SN(x_i)$ and remote point $x_i$ belongs to the $k_i$ nearest neighbors of point $x_j$. The equation is as follows:

$$(\exists x_j)(R_i = 1) \wedge (x_j \in SN(x_i)) \rightarrow x_i \in NN_{k_i}(x_j) \tag{6}$$

where $k_i$ is the searching round of $x_i$ $(i = 1, 2, ..., m)$ increasing from 1. $m$ is the number of remote points. It is noteworthy that the maximum search round corresponding to different remote points may be different.

**Definition 10** (Global stable state). The global stable state of the searching process is that all remote points reach the local stable state. At the same time, each remote point $x_i$ corresponds to a search round number $k_i(i = 1, 2, ..., m)$.

**Definition 11** (Natural neighbor eigenvalue). When the searching algorithm reaches the global stable state, the natural neighbor eigenvalue value $\mu$ is equal to the maximum of the searching round list.

$$\mu \overset{\Delta}{=} max(\{k_i|(\forall x_i)(\exists x_j)(R_i = 1) \wedge \\ (x_j \in SN(x_i)) \rightarrow x_i \in NN_{k_i}(x_j) \}) \tag{7}$$

Note that the natural neighbor eigenvalue is denoted as $\mu$ in ASNN and the natural neighbor eigenvalue is denoted as *supk* in [25] of Algorithm 1

ASNN is an accelerated version of the natural neighbor search algorithm. ASNN first to identify remote points according to Definitions 5 to 8. The next step is to search for the global stable state described in Definition 10, that is, to find at least one RKNN for each remote point and record the natural neighbor eigenvalue $\mu$. Finally, the natural neighbors of each data object are determined directly according to $\mu$ by Eq. (1).

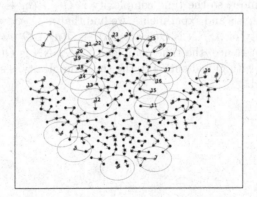

**Fig. 2.** Remote points and their remote neighborhood. (Color figure online)

Figure 2 visually shows the search results of ASNN in the synthetic dataset Flame. In Fig. 2, yellow circles indicate remote neighborhoods, and red points are remote points. The short gray line with an arrow indicates the nearest neighbor relationship between two data objects. As shown in Fig. 2, if the arrow points from $x_1$ to $x_2$, then $x_2$ is the nearest neighbor of $x_1$, $x_1$ is the reverse nearest neighbor of $x_2$, and the length of the arrow is the distance between the two points. $x_1$ and $x_2$ are the farthest nearest neighbors, so the remote radius $rd = d(x_1, x_2)$ can be found. Next, taking $x_3$ as an example, according to Definition 6 and Definition 7, the yellow circle is the remote neighborhood of $x_3$ and all black points in $S(x_3)$ are the remote region neighbors of $x_3$. Similarly, remote neighborhood and remote region neighbors of other data objects can be determined. $|SN(x_1)| = 1$, $|SN(x_2)| = 1$, $|SN(x_3)| = 3$, $|SN(x_4)| = 4$, $|SN(x_5)| = 4$, $|SN(x_6)| = 4$, $|SN(x_7)| = 4$. Although the size of remote neighborhoods is consistent, different data objects have a different number of remote region neighbors. The number of remote region neighbors of data objects in dense regions is significantly more than the number of remote region neighbors of points in sparse regions. Then $m$ remote points ($m << n$) are determined according to Definition 8. Searching the global stable state is the next step. Finally, the natural neighbors of each point can be found by the natural stable value $\mu$ at Line 32 of Algorithm 2. More details of ASNN are described in the Algorithm 2.

## 3.3   Complexity Analysis

Suppose there are $n$ points and $m$ remote points. The ASNN algorithm is composed of three main steps: (a) determine remote radius; (b) find remote points; (c) search reverse neighbors for remote points. With the help of $kd$-tree, step (a) can be completed in $O(n * log(n))$. Step (b) requires traversing the data set to determine whether the data points are remote or not, so the time complexity of step (b) is $O(n)$. In step (c), it is necessary to search for $\lambda$ neighbors of $m$ remote points at most, so the time complexity is $O(\mu * (m + \lambda))$. According to our theoretical analysis and experiments, we find that $m, \lambda << n$. Consequently, the time complexity of ASNN is $O(n * log(n)) + O(n) + O(\mu * (m + \lambda))$, where $\mu$, $m$ and $\lambda << n$. Hence, the time complexity of ASNN is $O(n * log(n))$, and in the worst case (with global outliers), the time complexity is similar to that of traditional method.

---

**Algorithm 2.** Accelerated Search for NaNs (ASNN)

---

**Input:** Dataset $X = \{x_1, x_2, \cdots, x_n\} \in \mathbb{R}$.
**Output:** Natural neighbor $NaN$, Natural Eigenvalue $\mu$, the number of reverse neighbors $nb$.
1: Initializing: $r \leftarrow 1$, $k_i \leftarrow 0$, $RS \leftarrow \emptyset$, $InS \leftarrow \emptyset$, $\mu \leftarrow 1$.
2: $n \leftarrow$ the total number of data points in $X$;
3: $Tree \leftarrow KDTree(X)$;   //create a kd-tree.
4: Finding the 1 nearest neighbor $NN_1(x_i)$ for each data object $x_i$ by $Tree$;
5: Using equation (3) to compute $rd$ ;   //determine remote radius.
6: //find remote region neighbors.
7: **for** each data object $x_i \in X$ **do**
8:    $SN \leftarrow$ find remote region neighbors by Definition 6 and Definition 7;
9:    $\overline{SN} \leftarrow$ the average number of the remote region neighbors of all data;
10: **end for**
11: //extract remote point
12: **for** each data object $x_i \in X$ **do**
13:    **if** $|SN(x_i)| < \overline{SN}$ **then**
14:       $RS \leftarrow RS \cup \{x_i\}$;   //$RS$ is the set of remote points
15:       $R_i \leftarrow 1$;
16:    **end if**
17: **end for**
18: //determine $\mu$
19: **while** true **do**
20:    $nbrs \leftarrow$ find the $r$-th nearest neighbor of each data object which belongs to the remote region neighbors of $RS$;
21:    $InS \leftarrow RS \cap nbrs$;
22:    **for** each data object $x_i \in InS$ **do**
23:       $RS - x_i$;
24:    **end for**
25:    **if** $RS$ is empty **then**
26:       break;
27:    **else**
28:       $r \leftarrow r + 1$;
29:    **end if**
30: **end while**
31: $\mu \leftarrow r$;
32: $NaN, nb \leftarrow$ two variables are obtained according to $r$ through equation (2);
33: return $NaN(x_i), nb, \mu$.

---

**Table 1.** The comparison of several different natural neighbors searching algorithms

|                       | NaNS   | kd-NaN            | FSNN       | ASNN       |
| --------------------- | ------ | ----------------- | ---------- | ---------- |
| Improved method       | –      | Indexing Method   | Find $supk$ | Find $supk$ |
| Searching scope       | Global | Global            | Global     | Local      |
| Distance matrix       | Yes    | No                | Yes        | No         |
| Single remote point   | –      | –                 | Yes        | No         |
| Time complexity       | $O(n^2)$ | $O(n * log(n))$ | $O(n^2)$   | $O(n * log(n))$ |

## 4 Experimental Evaluation

The running time of ASNN is compared with other natural neighbor searching algorithms on 15 synthetic datasets and 15 real-world datasets. The benchmarks include three searching algorithms for Natural Neighbors: (a) natural neighbor search algorithm based on distance matrix (abbreviated as NaNS), described in Algorithm 1, (b) natural neighbor search algorithm based on $kd$-tree [25] (kd-NaN for short), this algorithm is an improved version of NaNS by replacing step 3 in Algorithm 1 with $kd$-tree (using $kd$-tree to search for KNN and RKNN), and (c) fast search for natural neighbor algorithm FSNN. The time complexity of the above algorithms is listed in Table 1. All experiments are run on a PC with an AMD R7 37000X, 8G memory, 3.60GHz CPU, Windows 10, and Python 3.8.

### 4.1 Results on Synthetic Datasets

The comparative experiments of ASNN and other natural neighbor searching algorithms are executed on synthetic datasets in this section. Table 2 shows the adopted synthetic datasets with different sample numbers, dimensions, and categories.

Table 3 shows the average running time of 10 executions of comparative algorithms on 15 synthetic datasets. The black bold indicates that the algorithm takes the least time and has the fastest search speed.

In Table 3, ASNN is the fastest natural neighbor search algorithm, while NaNS is the slowest. kd-NaN and FSNN are similar. For each algorithm, the larger the number of samples, the more time the search usually takes. The shortest running time of NaNS among 15 synthetic datasets is Donut1 at 0.8999 s, and the most time-consuming is 134.3255 s on Cluto-t8-8k. NaNS is always the slowest search method on all datasets and takes the longest time. FSNN outperforms kd-NaN on 11 datasets, but slightly underperform on the Cure-t0-2000n-2D, Complex8, Complex9, and S-set2 datasets. FSNN focuses on the fast determination of the natural neighbor eigenvalue. Although the time complexity of FSNN is the same as NaNS, the distance matrix is searched only once in FSNN and the distance matrix is searched multiple times in NaNs. Therefore, FSNN's search takes less time than NaNS, especially in scenarios with a large natural

**Table 2.** Characteristics of 15 synthetic datasets.

| Datasets | Instances | Dimensions | Categories | Source |
|---|---|---|---|---|
| Aml28 | 804 | 2 | 5 | [2] |
| Donut1 | 1000 | 2 | 3 | [2] |
| Donutcurves | 1000 | 2 | 4 | [2] |
| Twenty | 1000 | 2 | 20 | [2] |
| Triangle2 | 1000 | 2 | 4 | [2] |
| Twomoons | 1502 | 2 | 2 | [2] |
| Cure-t0-2000n-2D | 2000 | 2 | 3 | [2] |
| Complex8 | 2551 | 2 | 8 | [2] |
| Complex9 | 3031 | 2 | 9 | [2] |
| D31 | 3100 | 2 | 31 | [2] |
| Dpb | 4000 | 2 | 5 | [2] |
| Banana | 4811 | 2 | 2 | [2] |
| S-set1 | 5000 | 2 | 15 | [2] |
| S-set2 | 5000 | 2 | 15 | [2] |
| Cluto-t8-8k | 8000 | 2 | 9 | [2] |

**Table 3.** Running time on 15 synthetic datasets (seconds).

| Datasets | NaNS | kd-NaN | FSNN | ASNN |
|---|---|---|---|---|
| Aml28 | 7.3023 | 3.1850 | 2.2213 | **0.5035** |
| Donut1 | 0.8999 | 0.1667 | 0.1526 | **0.0674** |
| Donutcurves | 1.5967 | 0.4101 | 0.2267 | **0.0696** |
| Twenty | 1.1308 | 0.2361 | 0.2153 | **0.0689** |
| Triangle2 | 1.2452 | 0.2090 | 0.1695 | **0.0749** |
| Twomoons | 4.2459 | 0.8110 | 0.4074 | **0.1947** |
| Cure-t0-2000n-2D | 2.6783 | 0.3011 | 0.4712 | **0.1581** |
| Complex8 | 4.5635 | 0.5437 | 0.7120 | **0.2190** |
| Complex9 | 6.1284 | 0.6540 | 1.0350 | **0.3018** |
| D31 | 36.5331 | 6.0579 | 2.4969 | **0.2730** |
| Dpb | 25.5979 | 2.4063 | 2.1519 | **0.6457** |
| Banana | 59.2210 | 7.5604 | 3.4572 | **0.5459** |
| S-set1 | 37.0130 | 3.4486 | 3.0556 | **0.6439** |
| S-set2 | 19.7640 | 1.8275 | 2.8970 | **0.5507** |
| Cluto-t8-8k | 134.3255 | 9.1865 | 8.7361 | **1.1398** |

neighbor eigenvalue. kd-NaN concentrates on changing the indexing method to reduce the time complexity to $O(supk * n * log(n))$. But it requires $supk$ times of global neighbor search. ASNN with the time complexity $O(n * log(n))$ only needs a global search once in finding remote points. Table 3 proves that ASNN outperforms other algorithms on 15 datasets.

In addition, we also demonstrate the consistency of the natural neighbor eigenvalue $\mu$ in ASNN and the natural eigenvalue $supk$ in NaNS [25]. Figure 3 visually shows the comparison results of the NaNS and ASNN for natural neighbor eigenvalue on 15 datasets. As shown in Fig. 3, the natural neighbor eigenvalue of the two algorithms is the same on 13 datasets out of 15 datasets. However, it is easy to find that the natural neighbor eigenvalue of ASNN is slightly larger than that of NaNS on dataset Dpb and dataset Banana, but the difference is not significant. A larger natural neighbor eigenvalue means that each sample has more natural neighbors, however, the difference between samples in sparse and dense regions does not disappear due to the change in the natural neighbor eigenvalue. In addition, it can be found from Table 3 that the improvement of the search efficiency of ASNN on the datasets Dpb and Banana is significant. In summary, ASNN is an effective natural neighbor search algorithm.

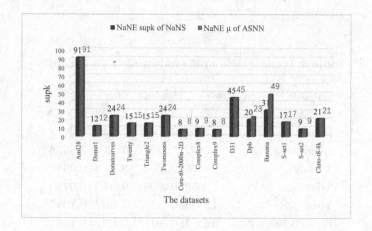

**Fig. 3.** Comparison of the natural neighbor eigenvalue on 15 datasets

## 4.2   Results on Real-World Datasets

To further verify the efficiency of ASNN, we conduct experiments on 15 real-world datasets. The characteristics of the experimental datasets are listed in Table 4. Average the running time of 10 executions of each algorithm is shown in Table 5.

Obviously, kd-NaN spends more time than FSNN on most datasets but runs faster than FSNN on Outlier and Optdigits. On data sets with fewer samples, such as Iris, Glass, the time-consuming of the four search algorithms is similar, but ASNN is more effective than NaNS, kd-NaN and FSNN. When the number and dimensions of the data are large, such as the Spambase dataset, all algorithms are time-consuming. NaNS, kd-NaN and FSNN all need thousands of seconds to reach a stable state of search, but ASNN only needs nearly 10 s to complete. On other large datasets, ASNN has obvious advantages. For example, on the Mfeat-fourier dataset, NaNS takes nearly 133 times that of ASNN, the search time of kd-NaN is 117 times that of ASNN, and FSNN takes about 47 times longer than ASNN. Similarly, on the Waveform dataset and Optdigits dataset, the advantage of ASNN is also obvious. Table 5 also proves that NaNS is usually the most time-consuming. ASNN requires less time than other algorithms and gets better results on all datasets. Ultimately, we can conclude that ASNN outperforms comparative algorithms in the average running time.

**Table 4.** Characteristics of 15 real-world datasets.

| Datasets | Instances | Dimensions | Categories | Source |
|---|---|---|---|---|
| Iris | 150 | 5 | 3 | [2] |
| Glass | 214 | 10 | 6 | [2] |
| Statlog (heart) | 270 | 14 | 2 | [2] |
| Ecoli | 336 | 8 | 8 | [2] |
| Liver-disorders | 345 | 7 | 2 | [2] |
| Ionosphere | 351 | 35 | 2 | [2] |
| Dermatology | 358 | 35 | 6 | [2] |
| Libras movement | 360 | 91 | 15 | [2] |
| Kdd-synthetic-control | 600 | 62 | 6 | [2] |
| DS1 | 699 | 11 | 2 | [2] |
| Outlier | 1641 | 3 | 1 | [2] |
| Mfeat-fourier | 2000 | 77 | 10 | [2] |
| Spambase | 4601 | 58 | 2 | [2] |
| Waveform | 5000 | 22 | 3 | [2] |
| Optdigits | 5620 | 3 | 10 | [2] |

According to the above experimental figures and analysis, ASNN has significantly higher search efficiency and faster search time than other natural neighbor search algorithms. The main reasons for the high efficiency of ASNN algorithm can be summarized into three aspects: (a) Fewer query points. ASNN determines natural neighbor eigenvalue by remote points instead of all samples, which greatly reduces the number of query points. (b) Smaller neighbor search range. ASNN searches for neighbors in the remote neighborhood instead of all samples. (c) Fewer iterations. ASNN quickly determines Natural neighbor Eigenvalue, avoids incremental neighbor search time and is a new effective reverse and local search strategy.

**Table 5.** Running time on 15 real-world datasets (seconds).

| Datasets | NaNS | kd-NaN | FSNN | ASNN |
|----------|------|--------|------|------|
| Iris | 0.0428 | 0.0234 | 0.0163 | **0.0110** |
| Glass | 0.0734 | 0.0253 | 0.0188 | **0.0136** |
| Statlog (heart) | 1.0274 | 0.7266 | 0.5204 | **0.1477** |
| Ecoli | 0.4899 | 0.2440 | 0.1571 | **0.0204** |
| Liver-disorders | 1.3273 | 0.7821 | 0.5825 | **0.0237** |
| Ionosphere | 10.3745 | 8.6183 | 6.9672 | **0.4117** |
| Dermatology | 0.3666 | 0.1930 | 0.0814 | **0.0511** |
| Libras movement | 0.2049 | 0.2153 | 0.0699 | **0.0364** |
| Kdd-synthetic-control | 0.4398 | 0.3133 | 0.0678 | **0.0673** |
| DS1 | 0.3717 | 0.1029 | 0.0649 | **0.0492** |
| Outlier | 1.6819 | 0.2149 | 0.2965 | **0.1310** |
| Mfeat-fourier | 80.5117 | 70.5958 | 28.5242 | **0.6032** |
| Spambase | 7515.3562 | 6922.7280 | 5943.5029 | **10.8404** |
| Waveform | 495.9092 | 175.5792 | 88.4808 | **5.8431** |
| Optdigits | 3492.7428 | 966.9830 | 2323.6464 | **4.5510** |

# 5   Conclusions

Natural neighbor is an effective neighbor concept and is widely praised for its application in many fields. However, the earlier algorithms for searching NaN traverse the entire dataset iteratively, which increases the unnecessary consumption of time on non-critical points, especially on large datasets. Inspired by sociology theory, an accelerated algorithm for searching natural neighbor ASNN is proposed to reduce the time-consuming of searching natural neighbor and further enhance the adaptability of natural neighbors in big data scenarios. ASNN can quickly determine natural neighbor eigenvalue based on the new concept remote points instead of all data objects. Hence, ASNN shortens the time of neighbor search without predetermining the number of neighbors. We attribute the better performance of ASNN to three aspects: fewer query points, local search strategy and fewer iterations. Numerous experiments have demonstrated that ASNN is faster and less time-consuming than NaNs [26], kd-NaN [25], FSNN [22].

In the future, we will abstract a natural neighbor search framework to explore the impact of each part on the search efficiency of natural neighbors. We are also interested in applying ASNN in others fields such as clustering analysis, outlier detection, instance reduction, etc.

**Acknowledgements.** This work is supported in part by National Natural Science Foundation of China under Grant 62006029, in part by Postdoctoral Innovative Talent Support Program of Chongqing under Grant CQBX2021024, in part by Natural Science Foundation of Chongqing (China) under Grant cstc2019jcyj-msxmX0683, cstc2020jscxlyjsAX0008, and in part by Project of Chongqing Municipal Education Commission, China under Grant KJQN202001434.

# References

1. Abu Alfeilat, H.A., et al.: Effects of distance measure choice on k-nearest neighbor classifier performance: a review. Big Data **7**(4), 221–248 (2019)
2. Asuncion, A., Newman, D.: UCI machine learning repository (2007)
3. Chen, Y.: Fast density peak clustering for large scale data based on kNN. Knowl.-Based Syst. **187**, 104824 (2020)
4. Chen, Y.: KNN-BLOCK DBSCAN: fast clustering for large-scale data. IEEE Trans. Syst. Man Cybern. Syst. **51**(6), 3939–3953 (2021). https://doi.org/10.1109/TSMC.2019.2956527
5. Cheng, D., Huang, J., Zhang, S., Zhang, X., Luo, X.: A novel approximate spectral clustering algorithm with dense cores and density peaks. IEEE Trans. Syst. Man Cybern. Syst. **52**(4), 2348–2360 (2022). https://doi.org/10.1109/TSMC.2021.3049490
6. Cheng, D., Zhang, S., Huang, J.: Dense members of local cores-based density peaks clustering algorithm. Knowl.-Based Syst. **193**, 105454 (2020)
7. Cheng, D., Zhu, Q., Huang, J., Wu, Q., Yang, L.: A novel cluster validity index based on local cores. IEEE Trans. Neural Netw. Learn. Syst. **30**(4), 985–999 (2019)
8. Cheng, D., Zhu, Q., Huang, J., Wu, Q., Yang, L.: Clustering with local density peaks-based minimum spanning tree. IEEE Trans. Knowl. Data Eng. **33**(2), 374–387 (2021)
9. Ester, M., Kriegel, H.P., Sander, J., Xu, X.: A density-based algorithm for discovering clusters in large spatial databases with noise. In: KDD, vol. 96, pp. 226–231 (1996)
10. Huang, J., Zhu, Q., Yang, L., Cheng, D., Wu, Q.: A novel outlier cluster detection algorithm without top-n parameter. Knowl.-Based Syst. **121**, 32–40 (2017)
11. Huang, J., Zhu, Q., Yang, L., Feng, J.: A non-parameter outlier detection algorithm based on natural neighbor. Knowl.-Based Syst. **92**, 71–77 (2016)
12. Jiang, A., Liu, J., Zhou, J., Zhang, M.: Skeleton extraction from point clouds of trees with complex branches via graph contraction. Vis. Comput. **37**(8), 2235–2251 (2021). https://doi.org/10.1007/s00371-020-01983-6
13. Li, J., Zhu, Q., Wu, Q., Fan, Z.: A novel oversampling technique for class-imbalanced learning based on smote and natural neighbors. Inf. Sci. **565**, 438–455 (2021)
14. Li, J., et al.: SMOTE-NaN-DE: addressing the noisy and borderline examples problem in imbalanced classification by natural neighbors and differential evolution. Knowl.-Based Syst. **223**, 107056 (2021)
15. Man, L., Mamoulis, N.: Reverse nearest neighbors search in ad-hoc subspaces. IEEE Trans. Knowl. Data Eng. **19**(3), 412–426 (2007)
16. Rodriguez, A., Laio, A.: Clustering by fast search and find of density peaks. Science **344**(6191), 1492–1496 (2014)
17. Srinilta, C., Kanharattanachai, S.: Application of natural neighbor-based algorithm on oversampling smote algorithms. In: 2021 7th International Conference on Engineering, Applied Sciences and Technology (ICEAST), pp. 217–220. IEEE (2021)
18. Stevens, S.S.: Mathematics, measurement, and psychophysics (1951)

19. Wahid, A., Annavarapu, C.S.R.: NaNOD: a natural neighbour-based outlier detection algorithm. Neural Comput. Appl. **33**(6), 2107–2123 (2021). https://doi.org/10.1007/s00521-020-05068-2
20. Wu, Z., Zeng, Y., Li, D., Liu, J., Feng, L.: High-volume point cloud data simplification based on decomposed graph filtering. Autom. Constr. **129**, 103815 (2021)
21. Yang, L., Zhu, Q., Huang, J., Cheng, D., Wu, Q., Hong, X.: Natural neighborhood graph-based instance reduction algorithm without parameters. Appl. Soft Comput. **70**, 279–287 (2018)
22. Yuan, M., Zhu, Q.: Spectral clustering algorithm based on fast search of natural neighbors. IEEE Access **8**, 67277–67288 (2020)
23. Zhang, Y., Ding, S., Wang, Y., Hou, H.: Chameleon algorithm based on improved natural neighbor graph generating sub-clusters. Appl. Intell. **51**(11), 8399–8415 (2021). https://doi.org/10.1007/s10489-021-02389-0
24. Zhao, S., Li, J.: A semi-supervised self-training method based on density peaks and natural neighbors. J. Ambient Intell. Humaniz. Comput. **12**(2), 2939–2953 (2021). https://doi.org/10.1007/s12652-020-02451-8
25. Zhu, Q., Feng, J., Huang, J.: Natural neighbor: a self-adaptive neighborhood method without parameter K. Pattern Recogn. Lett. **80**, 30–36 (2016)
26. Zou, X.L., Zhu, Q.S., Yang, R.L.: Natural nearest neighbor for Isomap algorithm without free-parameter. In: Advanced Materials Research, vol. 219, pp. 994–998. Trans Tech Publications (2011)

# A Recommendation Algorithm for Auto Parts Based on Knowledge Graph and Convolutional Neural Network

Junli Lin, Shiqun Yin[✉], Baolin Jia, and Ningchao Wang

Faculty of Computer and Information Science, Southwest University, Chongqing 400715, China
qqqq-qiong@163.com

**Abstract.** Due to the large number and complex types of auto parts, the ability that algorithm can accurately match the right auto parts is a major problem for buyers to solve. To address this problem, we propose a knowledge graph convolutional network with user history and item entity augmentation for auto parts recommendation system. First, the knowledge graph of auto parts and the knowledge graph of users to find a set of node examples. Second, collect contextual data about each instance. This data is to obtain information such as the hierarchical type, role, attribute value, and inferred neighbors of each neighbor instance from the knowledge graph. Finally, process this information into an array and use this array as the input to the neural network. After one or more aggregations and merging, as a vector of construction examples, the vector contains a wide range of information. The algorithm combines the advantages of knowledge graph and graph convolution, and combines with the idea of recommendation based on item content and recommendation based on user, so that the recommendation effect is improved.

**Keywords:** Knowledge graph (KG) · Graph Convolutional Network (GCN) · Recommendation · Data mining · Auto-parts

## 1 Introduction

The automobile industry [1, 2] is one of the largest industries in the world. The birth of a car often requires the joint production of many enterprises. The most basic part of automobile manufacturing is the parts required for automobiles, which are numerous and complex in type. There are many manufacturers of auto parts. And each manufacturer has different production levels and varieties. According to statistics, the number of OEMs in the industry accounts for less than 1%, and the dominant position of auto parts companies is obvious. Faced with the situation that there are fewer OEMs and more auto parts companies, buyers and auto parts suppliers have different problems. For buyers, due to geographical restrictions, vehicle sales in the peak season, and the situation that the demand for parts exceeds the supply, the supply of parts is at risk, and the purchaser cannot find replacement auto parts in time. For auto parts suppliers, there may be multiple

© The Author(s), under exclusive license to Springer Nature Singapore Pte Ltd. 2022
T. Li et al. (Eds.): BigData 2022, CCIS 1709, pp. 57–71, 2022.
https://doi.org/10.1007/978-981-19-8331-3_4

resale sales, resulting in the sales amount being much lower than the price given by the demand unit, and the profit is lower.

With the development of e-commerce platforms and the improvement of software and hardware performance, people's demand for obtaining accurate data is increasing day by day. In addition, the traditional collaborative filtering [3−8] algorithm is gradually moving closer to deep neural network (DNN) [9], knowledge graph (KG) [10−14] and other algorithms. In the paper, the recommendation model use KGCN_UI (Knowledge Graph Convolutional Networks with User History and Item Entities Augmentation) to enable users to predict parts, which is improved according to KGCN (Knowledge Graph Convolutional Networks for Recommender Systems) [12]. KGCN_UI combines knowledge graph and graph convolutional neural network, which can well solve the problem of data sparse and cold start. We build the information of parts into a knowledge graph. The knowledge graph can fully mine the rich relationship between entities, and can find the potential interests of users, with good diversity. Convolutional neural networks [15−21] can effectively extract data features, which are very important for the fields of computer vision and natural language processing. The data is not only regular network data, but also graph signal data. In this paper, we employ GCN to capture semantic information in knowledge graphs.

At present, the recommendation algorithms are mainly item-based recommendation or user-based recommendation [3−4]. User-based recommendation [5] is to use the current user's preference for items to find neighbors who have the same preferences as the current user, and then recommend the things that neighbor users like to the current user. Item-based recommendation is to find similar items based on the current user's preference for items, and then recommend similar items according to the user's historical preferences. This paper will combine the ideas of item-based recommendation and user-based recommendation to design, and improve the recommendation effect through their different advantages and disadvantages.

In fact, we use three algorithms to verify the effect of auto parts recommendation. The three algorithms are SVD [6], RippleNet [14] and KGCN [12]. The experimental results show that the accuracy of their recommendations is improved by 8.1%, 2.7% and 1.1% respectively.

Our contributions in this paper are summarized as follows:

- With KG as the carrier, the relationship between entities and entities is stored through graph data, and the recommendation model combines the advantages of KG and graph convolution to make the recommendation effect better.
- The idea of user − based recommendation and item − based recommendation is combined into the model to make its recommendation results more efficient and accurate.

## 2  Related Work

The traditional recommendation algorithm, collaborative filtering, has data sparseness and cold start problems. To solve these problems, many algorithms have been proposed. The matrix factorization algorithm [6−8] combines the characteristics of latent semantics and machine learning, and can mine deeper connections between users and items.

However, it has poor interpretability and generalization. Covington P et al. [9] applied DNN to recommender systems. Deep neural network has nonlinear fitting ability and good performance, but it cannot fully mine the relationship between entities. To solve such problems, in 2012, Google released the KG [22–23] for improving the quality of search. Therefore, many people who develop recommendation algorithms turn their attention to knowledge graphs. RippleNet [14] is a memory network–like model that propagates users' latent preferences in KG and explores their hierarchical interests. The results recommended by KGCN [12] are very good in diversity and accuracy. However, it should be noted that in RippleNet, the importance of relationships is very weak; in KGCN, it pays more attention to the expansion of item entities, and it is less used for user history. As the saying goes, "Things gather together, people are divided into groups". Friends influence user choices. KGCN_UI adds social network information to KGCN, that is, the influence information of friends. This method can effectively improve the accuracy and user satisfaction.

In order to better extract the features of graph data, many scholars have made improvements based on convolutional neural networks and proposed graph convolutional neural network algorithms. In 2013, Bruna et al. presented the first significant study on GCNs, developing a variant of graph convolution [16–18] based on spectrogram theory. Since spectral methods usually process the entire graph at the same time, and it is difficult to parallelize or scale to large graphs, spatial-based graph convolutional networks [19–21] have been rapidly developed. Together with the sampling strategy, it can be computed in a batch of nodes instead of the entire graph, improving efficiency. Ge, Y et al. [24] proposed a new recommendation algorithm based on graph convolutional networks, using two sets of graph convolution operations to utilize two different interactive information simultaneously. It provides us with an idea to apply the idea of user-based recommendation and content-based recommendation to the recommendation model.

## 3 Proposed Method

### 3.1 Knowledge Graph

The knowledge graph is to build the relationship between entities, in which the triplet of "entity-relationship-entity" is the constituent unit of the knowledge graph. Moreover, the entities can be connected to each other through the relationship to form a network knowledge structure.

The KG of the commodities $G_v(h, r, t)$, where $h \in \varepsilon$, $r \epsilon R$, $t \in \varepsilon$; $\varepsilon$ represents the entity set in the knowledge graph; R represents the relation set in the knowledge graph. For example, $G_v(bolt, supplier, ACompany)$ means that A Company produced the bolts (Fig. 1).

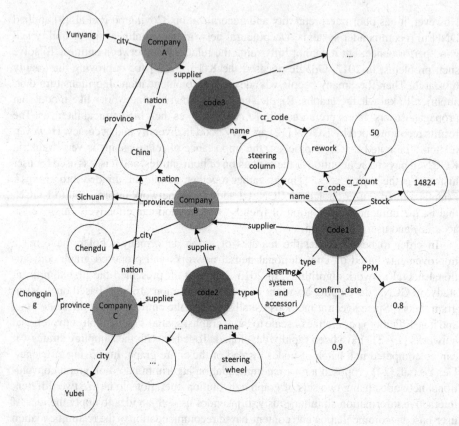

**Fig. 1.** The knowledge graph of parts. It includes information about manufacturer, name, type, confirm rate, Part Per Million (PPM), failure record (cr_count, cr_code), stock, address, etc. link: https://pan.baidu.com/s/1yagdgyZveR30umIlktpTCg?pwd=1234

The KG of the user $G_u(u, d, f)$ is mainly constructed based on the similarity of the user's purchase records and the user's basic information. Where $u \in \tau$, $d \in D$, $f \in \tau$; $\tau$ represents the set of user entities in the knowledge graph; D represents the relational set of user KG in the knowledge graph. For example, $G_v(APurchasingDepartmen, city, Chengdu)$ means that A Purchasing Department is located in Chengdu (Fig. 2).

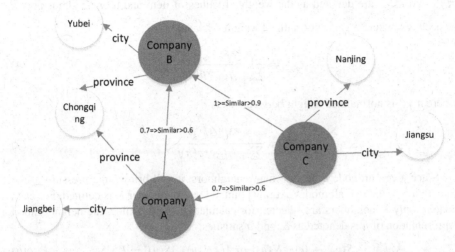

**Fig. 2.** The knowledge graph of users. It includes information about users' friends, address, etc. link: https://pan.baidu.com/s/1yagdgyZveR30umIlktpTCg?pwd=1234

### 3.2  KGCN_UI Layer

Existing user-item interaction matrix $\in R^{M \times N}$, is defined according to users' implicit feedback, where $M$ represents the number of users and $N$ represents the number of items.

The central idea of KGCN_UI is to use the message passing mechanism of graph neural network to combine training with basic recommendation ideas. KGCN_UI can extract effective information in $G_u$ and $G_v$, predict the user's preference for items, and then use the interaction matrix $Y$ to train the parameters to obtain a better prediction effect.

$G_v$ and $G_u$ are weighted graphs, and the relationship becomes a weight. This weight can be understand as the preference degree that the relationship affects the user's behavior.

$$\pi_r^u = g(u, r) \tag{1}$$

where $u$ represents the user vector, $r$ represents the relationship set vector. It represents the importance of relation vector $r$ to user vector $u$, and the result can better reflect the attention of user vector $u$ than directly removing the relation vector $r$.

$$\pi_d^v = g(v, d) \tag{2}$$

where $\pi_d^v$ is $G_u$'s weight, $v$ is the item vector, $d$ is relation between users.

In order to pass the surrounding node information to the central node, the weighted summation of adjacent nodes is required:

$$v_{N(v)}^u = \sum_{e \in N(v)} \tilde{\pi}_{r_{v,e}}^u e, \quad u_{N(u)}^v = \sum_{f \in N(u)} \tilde{\pi}_{r_{u,f}}^v f \tag{3}$$

$v_{N(v)}^u$ and $u_{N(u)}^v$ are denoted as the weighted values of item and user neighbor nodes, respectively, where $\widetilde{\pi}_{r_{v,e}}^u$ is normalized weight between item

$$\widetilde{\pi}_{r_{v,e}}^u = \frac{exp(\pi_{r_{v,e}}^u)}{\sum_{e \in N(V)} exp(\pi_{r_{v,e}}^u)}, \tag{4}$$

where $\widetilde{\pi}_{d_{u,f}}^v$ is normalized weight between users

$$\widetilde{\pi}_{d_{u,f}}^v = \frac{exp(\pi_{d_{u,f}}^v)}{\sum_{f \in F(U)} exp(\pi_{d_{u,f}}^v)} \tag{5}$$

Since a certain node $n$ has too many neighbors, it will bring huge pressure to the calculation of the overall model. At this point, a hyperparameter $K$ is defined. For each node $n$, only $K$ neighbors are selected for calculation. At this time, the neighborhood representation of $n$ is denoted as $S$, and it satisfies:

$$S(n) \rightarrow \{e|e \in N(n)\} \ or \ \{f \in N(u)\}, |S(n)| = K \tag{6}$$

Finally, it requires a message aggregation, that is, one or more operations of the fully connected layer.

$$v' = \sigma_v(W_v \cdot agg(e, v_{S(v)}^u) + b_v) \tag{7}$$

where $\sigma_v$ is the nonlinear activation function, such as Relu, sigmoid, $W_v$ is the linear transformation matrix, and $b_v$ is the bias term, $agg(v, v_{S(v)}^u)$ means that the message aggregation is performed on the item $v$ again, and $S(v)$ represents all the neighborhoods of $v$, which refers to the vector generated by the previous iteration of the item $v$.

There are three aggregation methods:

1. Summation aggregation:

$$agg_{sum} = e + v_{S(v)}^u \tag{8}$$

The corresponding element bits are added together.

2. Splicing aggregation:

$$agg_{concat} = concat(e, v_{S(v)}^u) \tag{9}$$

That is to concatenate the vector with the vector $e$. If the original dimensions are $F$, then the dimension of the spliced vector is $2F$, so the outer linear change matrix $W$ and the dimension of the bias term $\mathbf{b}$ also need to be changed accordingly.

3. Neighbor aggregation:

$$agg_{neighbor} = v_{S(v)}^u \tag{10}$$

It is directly used as the network output vector.

In the same way, the feature vector of the user also needs to be aggregated again:

$$u' = \sigma_u(W_u \cdot agg(f, u_{S(u)}^v) + b_u) \tag{11}$$

where $\sigma_u$ is the nonlinear activation function, $W_u$ is the linear transformation matrix, and $b_u$ is the bias term. In the formula, $agg(f, u_{S(u)}^v)$ means that the message aggregation is performed on the user $u$ again, and $S(u)$ represents all the neighborhoods of user $u$.

### 3.3 Learning Algorithm

Through KGCN_UI, it can be seen that the final result of the entity depends on its own information and the information of neighbor nodes. We first query the KG to find a set of example nodes. Subsequently, contextual data about each example is collected. We retrieve relevant data from the KG, including the hierarchical type, role, and attribute values of each neighbor instance encountered, as well as inferred neighbors. These data are processed into arrays as input to the neural network. Through aggregation and composition, we can build a single vector representation of the instance, which contains extensive contextual information (Fig. 3).

Then the predicted preference of user vector $u$ to item vector $v$ is

$$y' = predict(v, u', v', u) \tag{12}$$

where the user vector is u, and $u'$ represents the feature vector of the aggregated user vector $u$; the item vector is v, and $v'$ is represents the feature vector of the aggregated item vector $v$. And $predict()$ is an arbitrary function, such as an inner product. It is the predicted value, or the predicted click rate of user vector $u$ on item vector $v$

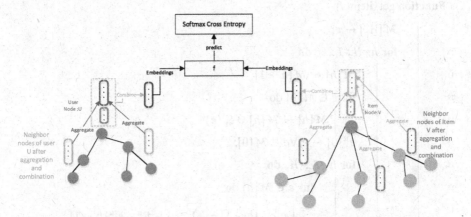

**Fig. 3.** Our KGCN_UI architecture. The figure contains the KG of the commodities, $G_u$ and The KG of the users, $G_v$. They all need to perform feature extraction to obtain feature vectors $u'$ and $v'$, and then calculate with user vector $u$ and item vector $v$ to obtain the predicted value $y'$.

---

**Algorithm 1**

---

**Input:**

Knowledge graph $G_v(\mathcal{E},\mathcal{R})$;User-to-user relationship graph $G_u(\mathcal{T},\mathcal{D})$;interaction matrix $Y$; neighborhood sampling mapping $\mathcal{S}_u, \mathcal{S}_v$; trainable parameters: $\{u\}_{u \in \mathcal{U}}$, $\{v\}_{v \in \mathcal{V}}$, $\{e\}_{e \in \mathcal{E}}$, $\{r\}_{r \in \mathcal{R}}$, $\{f\}_{f \in \mathcal{T}}$, $\{d\}_{d \in \mathcal{D}}$;hyper-parameters : $H_v$, $H_u$, $predict(\cdot)$, $agg(\cdot)$

**Output:**

Prediction function $\mathcal{F}(v,u)$

1 **while** KGCN_UI not converge **do**

2      **for** $(u,v)$ in $Y$ **do**

3          $v' \leftarrow get\_item(v)$

4          $u' \leftarrow get\_user(u)$

5          Calculate predicted probability $y'_{uv} = predict(v, u', v', u)$;

6          Update parameters by gradient descent;

7 **return** $\mathcal{F}$

8 **Function** get_item(v)

9      $\mathcal{M}[H_v] \leftarrow v$;

10      **for** $h = H_v\text{-}1,...,0$ **do**

11          $\mathcal{M}[h] \leftarrow \mathcal{M}[h+1]$;

12          for $e \in \mathcal{M}[h]$ do

13             $\mathcal{M}[h] \leftarrow \mathcal{M}[h] \cup \mathcal{S}_v(e)$

14      $e^u[0] \leftarrow e, \forall e \in \mathcal{M}[0]$;

15      **for** $h = 1,...,H_v$ **do**

16          **for** $e \in \mathcal{M}[h]$ **do**

17             $e^u_{\mathcal{S}_v(e)}[h-1] \leftarrow \sum_{e \in \mathcal{S}_v(e)} \tilde{\pi}^u_{r_{e,e'}} e'^u[h-1]$

18             $e^u[h] \leftarrow agg(e^u_{\mathcal{S}_v(e)}[h-1], e^u[h-1])$

19      $v' \leftarrow e^u[H_v]$

20 **Return** $v'$

21 **Function** $get\_user(u)$

22      $\mathcal{M}[H_u] \leftarrow u$;

23      **for** $h = H_u\text{-}1,...,0$ **do**

24          $\mathcal{M}[h] \leftarrow \mathcal{M}[h+1]$;

25          **for** $f \in \mathcal{M}[h+1]$ **do**

26    $\mathcal{M}[h] \leftarrow \mathcal{M}[h] \cup \mathcal{S}_u(f)$

27    $f^v[0] \leftarrow f, \forall f \in \mathcal{M}[0];$

28    **for** $h=1,...,H_u$ **do**

29        **for** $f \in \mathcal{M}[h]$ **do**

30            $f^v_{\mathcal{S}_u(f)}[h-1] \leftarrow \sum_{f \in \mathcal{S}_u(f)} \tilde{\pi}^v_{r_{f,f'}} f'^v[h-1]$

31            $f^v[h] \leftarrow agg(f^v_{\mathcal{S}_u(f)}[h-1], f^v[h-1])$

32    $u' \leftarrow f^v[\mathrm{H}_u]$

33 **return** $u'$

---

In order to speed up the training and reduce the number of updated weights, we will negatively sample the nodes. The loss function is as follows:

$$Loss = \sum_{u \in \mathcal{U}} (\sum_{v:y_{uv}=1} \mathcal{L}(y_{uv}, y'_{uv}) - \sum_{i=1}^{T^u} E_{v_i \sim P(v_i)} \mathcal{L}(y_{uv_i}, y'_{uv_i})) + \lambda ||\mathcal{F}||_2^2 \quad (13)$$

## 4  Experiment

In this section, we will adopt SVD, RippleNet, KGCN and our algorithm to compare the recommendation results of auto parts.

### 4.1  Datasets

In Sect. 3.1, the construction of the knowledge graph for auto parts is explained, which will not be repeated here. In addition, in order to verify whether KGCN_UI has a certain improvement in the recommendation of other datasets, we use the Last.FM and MovieLens dataset for verification.

**Table 1.** Basic statistics and hyper- parameter settings for the three datasets (K: neighbor sampling size, d: dimension of embeddings, H: depth of receptive field, $\lambda$: L2 regularizer weight, $\eta$: learning rate, user_entities: users and users' attribute information, Item_entities: items and items' attribute information).

|  | auto_part | Last.FM | MovieLens |
|---|---|---|---|
| Users | 36 | 1872 | 138159 |
| Items | 81944 | 3846 | 16954 |
| Interactions | 75222 | 72346 | 14947311 |
| Item_entities | 10934 | 9366 | 102569 |
| Item_relations | 11 | 60 | 32 |
| Item_KG_triples | 709899 | 15518 | 499474 |
| User_entities | 36 | 1872 | 2433 |
| User_relations | 12 | 50 | 50 |
| User_KG_triples | 1614 | 1907 | 169339 |
| K | 8 | 8 | 8 |
| d | 16 | 16 | 16 |
| H | 1 | 1 | 1 |
| $\lambda$ | 1e − 7 | 1e − 4 | 1e − 4 |
| $\eta$ | 5e − 3 | 5e − 4 | 5e − 4 |
| Batch size | 64 | 128 | 65536 |

## 4.2  Baselines

We use KGCN_UI on a dataset of auto-parts and compare with the following algorithms, where SVD is KG-free and RippleNet and KGCN are KG-aware methods. In addition, I also compared with different baselines for the lastFM dataset.

SVD is a matrix factorization technique in linear algebra that can decompose the user-item interaction matrix. RippleNet is a memory − network-like approach that propagates users' preferences on the KG for recommendation. KGCN is an end-to-end framework that captures inter-item relatedness effectively by mining their associated attributes on the KG.

**Table 2.** The results of AUC and F 1 in CTR prediction of different baselines.

| Heading level | AUC (auto-part) | F1 (auto-part) | AUC (lastFM) | F1 (lastFM) | AUC (MovieLens) | F1 (MovieLens) |
|---|---|---|---|---|---|---|
| SVD | 0.744 (−16%) | 0.666 (−23%) | 0.656 (−15.2%) | 0.655 (−15.3%) | 0.963 (−2.2%) | 0.919 (−2.6%) |
| RippleNet | 0.798 (−10.6%) | 0.774 (−12.2%) | 0.780 (−2.8%) | 0.702 (−10.6%) | 0.968 (−1.7%) | 0.912 (−3.3%) |
| KGCN | 0.880 (−2.4%) | 0.869 (−2.7%) | 0.796 (−1.2%) | 0.721 (−8.7%) | 0.977 (−0.8%) | 0.931 (−1.4%) |
| KGCN_UI(sum) | **0.904*** | **0.896*** | 0.795 (−0.7%) | 0.795 (−0.7%) | **0.985*** | 0.944 (−0.1%) |
| KGCN_UI(concat) | 0.890 (−1.4%) | 0.881 (−1.5%) | **0.808*** | **0.808*** | 0.983 (−0.2%) | 0.942 (−0.3%) |
| KGCN_UI(neighbor) | 0.895 (−0.9%) | 0.887 (−0.9%) | 0.798 (−1%) | 0.798 (−1%) | **0.985*** | **0.945*** |
| KGCN_UI(avg) | 0.896 (−0.8%) | 0.888 (−0.8%) | 0.800 (−0.8%) | 0.799 (−0.9%) | 0.984 (−0.1%) | 0.944 (−0.1%) |

## 4.3  Experiments Setup

In KGCN_UI, we set g as the inner product, $f$ as the average function, and σ as *ReLU* and *tanh*, which are used for non-last layer aggregators and last layer aggregators, respectively. Other parameters are shown in Table 1. The dataset is divided into training set, evaluation set and validation set with a ratio of 6:2:2. To validate algorithm performance, we use AUC and F1 to evaluate CTR predictions. All training parameters are optimized by Adam algorithm. Under python3.8, torch1.17.1 and numpy1.19.5, the encoding of KGCN_UI is implemented (Table 3).

The hyper-parameters for each baseline are as follows.

Other hyper-parameters are consistent with the papers corresponding to the respective baselines.

**Table 3.** The hyper – parameters of each baseline.

| Data | | SVD | RippleNet | KGCN |
|---|---|---|---|---|
| Auto – part | d | / | 16 | 16 |
| | H | / | 1 | 1 |
| | λ | 1e − 1 | 1e − 7 | 1e − 4 |
| | η | 4e − 2 | 2e − 2 | 5e − 4 |
| Last.FM | d | / | 16 | 16 |
| | H | / | 1 | 1 |
| | λ | 1.5e − 1 | 1e − 7 | 1e − 4 |
| | η | 4e − 1 | 2e − 2 | 5e − 4 |
| MovieLens | d | 8 | 8 | 32 |
| | H | / | 2 | 2 |
| | λ | / | 1e − 6 | 1e − 7 |
| | η | 0.5 | 1e − 2 | 2e − 2 |

### 4.4 Results

Based on the experiments, we draw the following conclusions:

1) From Table 2, we can see that the RippleNet, KGCN and KGCN_UI algorithms have better recommendation performance than SVD. Thus, it can be concluded that KG-aware methods are helpful for accurate recommendation.
2) By comparing the AUC and F1 of KGCN and KGCN_UI, it can be seen that the recommendation performance has been improved after combining the knowledge graph of the item and the knowledge graph of the user.
3) From Table 2, KGCN_UI is also applicable to other application scenarios.

**Impact of Neighbor Sampling Size**

Table 4. Impact of neighbor sampling size

| K | 2 | 4 | 8 | 16 |
|---|---|---|---|---|
| AUC | 0.835 | 0.869 | **0.904** | 0.876 |
| F1 | 0.819 | 0.858 | **0.896** | 0.865 |

We studied the Table 4, and it can be obtained from the table that when $k = 8$, the recommendation performance is the best. This is because k is too small to effectively obtain domain information. If k is too large, too much information is obtained and the noise is larger (Table 5).

**Impact of Depth of Receptive Field**

Table 5. Impact of depth of receptive field.

| H | 1 | 2 | 3 | 14 |
|---|---|---|---|---|
| AUC | **0.904** | 0.862 | 0.741 | 0.752 |
| F1 | **0.896** | 0.849 | 0.723 | 0.746 |

We change $h$ from 1 to 4 and get the best result when $h = 1$. Since the setting of $h$ is too large, on the one hand, it is easy to cause a large amount of calculation, and on the other hand, information will overlap.

**Impact of Dimension of Embedding**

Table 6. Impact of neighbor sampling size

| d | 4 | 8 | 16 | 32 |
|---|---|---|---|---|
| AUC | 0.836 | 0.887 | **0.904** | 0.899 |
| F1 | 0.819 | 0.876 | **0.896** | 0.893 |

From Table 6, we can know that when $d = 16$, the result is the best. The reason for this result is that the $d$ value is too small to be informative, and the $d$ value is too large to easily cause overfitting.

## 5 Conclusion

This paper proposes KGCN_UI. KGCN_UI uses the GCN method to extract the user's preference information for items in the KG. At the same time, this paper adopts the idea of two models of user-based recommendation and item-based recommendation, which makes the recommendation results more accurate. Through experiments on the data set of auto parts, it can be effectively proved that the KGCN-UI algorithm improves the accuracy of the recommendation results. In addition, we also compared other datasets, such as music and movie. From the experimental results, it can be concluded that KGCN-UI has surpassed the best baselines.

Our future work is mainly in two aspects.

1. Explore non-uniform samplers. Currently, we use uniform sampling to construct the receptive field of the entity. In the future, we will focus on non-uniform adopters for future work.
2. Data problems. The KG of the users is mainly constructed based on the similar reading of the user's purchase record and address. Later, we will consider collecting other information of the user, such as sales ability, etc. to add to the KG. In the future, the KG of auto parts can start from the supply chain and use real-time transportation conditions to improve the performance of recommendation.

**Acknowledgment.** This work is supported by the Science &Technology project (4411700474, 4411500476).

# References

1. Qiu, B., Wang, F., Liu, W.: Development trend of China's auto industry. Auto Ind. Res. **1**, 2–9 (2022)
2. Fu, G.: Analysis of the Automotive Industry. Tongji University Press, Tongji (2018)
3. Sarwar, B., Karypis, G., Konstan, J., Riedl, J.: Item_Based collaborative filtering recommendation algorithm. In: Proceedings of the 10th International World Wide Web Commerce. Hong Kong, pp. 285–288 (2001)
4. Balabanovic, M., Shoham, Y.: FAB: content–based collaborative recommendation. Commun. ACM **40**(3), 66–72 (1997)
5. Herlocker, J.L., Konstan, J.A., Borchers, A., Riedl, J.: An algorithmic framework for performing collaborative filtering. In: Proceedings of SIGI. Berkley, pp. 227–234 (1999)
6. Yehuda K.: Factorization meets the neighborhood: a multifaceted collaborative filtering model. In: Proceedings of the 14th ACM SIGKDD International Conference on Knowledge Discovery and Data Mining. New York, pp. 426–434 (2008)
7. Resnick, P., Iacovou, N., Suchak, Mi., Bergstrom, P. Riedl, J.: GroupLens: an open architecture for collaborative filtering of netnews. In: Proceedings of the 1994 ACM conference on Computer supported cooperative work (CSCW '94). Association for Computing Machinery, New York, pp. 175–186 (1994)
8. Koren, Y.: Collaborative filtering with temporal dynamics. In: Proceedings of the 15th ACM SIGKDD International Conference on Knowledge Discovery and Data Mining. Association for Computing Machinery, New York, pp. 447–456 (2009)
9. Covington, P., Adams, J., Sargin, E.: Deep neural networks for youtube recommendations. In: ACM Conference on Recommender Systems. New York, pp. 191–198 (2016)
10. Chen, Y., Feng, W., Huang, M., Feng, S.: Collaborative filtering recommendation algorithm of behavior route based on knowledge graph. Comput. Sci. **48**, 176–183 (2021)

11. Wang, H., Zhang, F., Xie X., Guo, M.: DKN: Deep knowledge–aware network for news recommendation. In: Proceedings of the 2018 World Wide Web Conference on World Wide Web. Geneva, pp. 1835–1844 (2018)

12. Wang, H., Zhao, M., Xie, X., Li, W., Guo, M.: Knowledge graph convolutional networks for recommender systems. In: Proceedings of the 2019 World Wide Web Conference. New York, p. 7 (2019)

13. Wang, X., He, X., Cao, Y., Liu, M., Chua, T.: KGAT: knowledge graph attention network for recommendation. In: Proceedings of KDD. Anchorage, pp. 4–8 (2019)

14. Wang, H., et al.: RippleNet: propagating user preferences on the knowledge graph for recommender systems. In: Proceedings of the 27th ACM International Conference on Information and Knowledge Management. ACM (2018)

15. Kipf, T.N., Welling, M.: Semi–supervised classification with graph convolutional networks. In: Proceedings of ICLR (2017)

16. Schlichtkrull, M., Kipf, T.N., Bloem, P., van den Berg, R., Titov, I., Welling, M.: Modeling relational data with graph convolutional networks. In: The Semantic Web. ESWC 2018. Lecture Notes in Computer Science, vol. 10843. Springer, Cham (2018). https://doi.org/10.1007/978-3-319-93417-4_38

17. Xu, B., Shen, H., Cao Q., Qiu, Y., Cheng, X.: Graph wavelet neural network. In: Proccessdings of ICLR (2019)

18. Zhuang, C., Ma, Q.: Dual graph convolutional networks for graph–based semi–supervised classification. In: Proceedings of WWW. Geneva, pp. 499–508 (2018)

19. Gao, H., Wang, Z.: Large–scale learnable graph convolutional networks. In: Proceedings of KDD. London, pp. 1416–1424 (2018)

20. Velickovic, P., Cucurull, G., Casanova, A., Romero, A., Lio, P., Bengio, Y.: Graph attention networks. In: Proceedings of ICLR. Vancouver (2018)

21. Hamilton, W.L., Ying, Z., Leskovec, J.: Inductive representation learning on large graph. In: Proceedings of NIPS. Los Angeles, pp. 1024–1034 (2017)

22. Amit, S.: Introducing the knowledge graph. America: Official Blog of Google (2012)

23. Xu, Z., Sheng, Y., He, L., Wang, Y.: Review on knowledge graph techniques. J. Univ. Electron. Sci. Technol. China **45**, 589–606 (2016)

24. Ge, Y., Chen, S.C.: Graph convolutional network for recommender systems. In: Ruan Jian Xue Bao/J. Sotfw. **31**, 1101–1112 (2020)

# Identifying Urban Functional Regions by LDA Topic Model with POI Data

Yuhao Huang[1,2], Lijun Zhang[1,2(✉)], Haijun Wang[1,2], and Siqi Wang[1,2]

[1] School of Computer Science, Northwestern Polytechnical University, Xi'an, China
zhanglijun@nwpu.edu.cn
[2] Key Laboratory of Big Data Storage and Management, Northwestern Polytechnical University, Ministry of Industry and Information Technology, Xi'an, China

**Abstract.** Identifying Urban Functional Regions (UFR) can achieve the rational aggregation of social resource space, realize urban economic and social functions, promote the deployment of urban infrastructure, radiate and drive the development of surrounding regions, so the identification of urban functional regions can promote the efficient development of cities. However, the traditional functional region identification method is mainly based on remote sensing mapping, which relies more on the natural geographical characteristics of the region to describe and identify the region, while the urban functional region is closely related to human activities, and the traditional functional region identification results are not ideal. Social data includes a series of data that reflect people's activities and behaviors, such as trajectory data, social media data, and travel data, thus the analysis of social data can more effectively solve the difficulties of traditional mapping and identification. POI (Point of Interest) data, as a typical type of social data, can be used to identify urban functional regions. We apply the LDA topic model to the POI data, and propose a new urban functional region identification method, which makes full use of the POI data to reflect the activity categories of urban populations to characterize the features of regional functions and achieve a high degree of identification of urban functional regions. Through experimental verification on real data, the experimental results show that the proposed method can more accurately identify urban functions, which proves the method reliable.

**Keywords:** Urban functional regions identification · POI · LDA topic model · Clustering algorithm

## 1 Introduction

The city consists of a variety of regions that provide different functions to meet the different needs of urban residents for food, clothing, housing and transportation [1,2]. The identification of urban functional regions(UFR) can provide reference for urban planning and make overall arrangements [3,4]. It can promote the rational organization of land use in all parts of the city, develop simultaneously, realize complementarity, and finally achieve an organic whole [5,6]. Starting from the agglomeration effect, the advantages of functional regions such as technology, management, concept and capital can be infiltrated into the surrounding regions to drive the development of the surrounding regions [7,8]. Therefore, the urban functional region is a regional space that can realize the spatial aggregation of relevant social resources, effectively play a specific urban function, and is the spatial carrier to realize the economic and social functions of the city, so it is of great significance to identify the urban functional regions.

For example, When you need to plan an industrial zone in the city, you should allocate the downwind side of the living and residential land according to the local dominant wind direction, and to meet requirements of water sanitation protection, the industrial land should be located downstream of the river water source. Similarly, residential areas should choose the best sanitary conditions in the city: high terrain, dry air, not threatened by flooding, clean soil or harmless pollution, close to the surface water, and more sunshine. If the planning is bad, lots of problems will arouse: transportation will be more complicated, the cost of operation and management will increase, and the urban environmental sanitation will be poor. So identifying urban functional areas is full of importance and values our studying.

Up to now, many researchers have deeply studied the division of urban functional regions. Traditionally, the use of remote sensing image data space mapping to identify the general urban functional regions, this approach will extract the types of surface coverage data at all levels in remote sensing image. And the spatial resolution of remote sensing data, spectral resolution and time resolution(three resolution metrics commonly used in remote sensing data) is improved due to the development of the existing sensor and data fusion technology, which can enhance precision and accuracy of remote sensing data and provide more useful information of the urban functional regions identification [9,10]. However, on the one hand, remote sensing data is limited by the sensor bottleneck, and it is difficult to accurately identify single remote sensing data, which has an impact on the classification of urban functional regions, and the acquisition of multiple remote sensing data is also expensive. On the other hand, remote sensing data only describe the natural characteristics of urban regions, and the functions of urban regions are often closely related to human activities [11].

Differently, various social data, such as trajectory data and media information, can be used to divide urban functional regions from the perspective of humanities and social sciences, providing meaningful advice for urban construction. Social data has the characteristics of massiveness, pluralism, strong humanistic attributes, and concentrated group effect, which can not only simplify data acquisition, but also excavate the formation and interaction law of group behavior, explore the mechanism of information dissemination and evolution. On this basis, policymakers can conduct in-depth analysis of the development and existing layout of urban functional areas from the perspective of public expectations. Zhang et al. used urban taxi track data combined with Voronoi diagram to identify urban functional regions, and researchers' flow is correlated with urban functional regions [12]. Volgmann et al. plotted the functional spatial distribution of German metropolis from the perspective of commercial finance, policy, media culture, historical background and other attributes based on the effect of "borrowing size", and obtained the special structure of multi-centralization [13]. Gao et al. integrated multi-feature potential semantic information of remote sensing image and POI data and embedded the topic model, which significantly improved the identification accuracy of urban functional regions [11].

Among them, POI data, as a kind of social data, is a feature point set containing spatial attribute information such as name, address, function, longitude and latitude, which contains rich cultural and economic features as well as natural features and can reveal the use function of urban land [14]. However, due to the complexity of urban regional functions, it is not accurate to identify urban functional regions only by analyzing the frequency of POI categories. Therefore, this paper will use the characteristics of LDA topic model to apply it to the embedded representation of POI data to identify urban functional regions. Based on the relationship and frequency characteristics of POI spatial locations, this method mines more semantic information of urban regions and improves the accuracy of functional region identification results.

The research in this paper contributes to the exploration of urban computing based on POI data, and helps to understand urban dynamics and realize smart cities. There are three main contributions of this paper:

(1) This paper proposes an embedded representation method that applies the topic model to blocks, which are divided according to a variety of Unit cut in grids of specified sizes. This approach treats blocks as documents, specific city functions as keywords, and POI within the Block region as words. In this way, we use the distribution of POI data to determine the thematic distribution of urban functions, and then through the thematic distribution, the UFR is represented. The model has strong adaptability and improves the spatial connection of data.

(2) Regions are divided by clustering method to identify UFR and infer specific functions. Each block is represented by a POI set, and we merge similar blocks into UFR, and different UFR represent different functions when we make sure blocks in different UFR will differ from each other. According to the UFR identifying results, the function label of the region is obtained by the weight of the inner topic and the attribute of the topic.

(3) This paper uses real geographical data to analyze the effect of functional region identification in detail, and verifies the feasibility of the POI data analysis method based on topic model. The data analysis not only prove the feasibility of the methods in this paper, but also help study the city development direction and analysis of construction objectives.

The rest of this paper is organized as follows: Sect. 2 analyzes the research progress and shortcomings of the existing urban functional regions, including the traditional and modern different two categories of methods. Section 3 gives a detailed description of the mining method of function distribution, including the embedding representation and clustering process of blocks. Section 4 analyzes some important parameters in this paper and obtains the optimal values, such as Block size and number of keywords. Section 5 gives a brief summary of this paper.

## 2   Related Works

Different from the traditional land use and land cover classification (LULC), the identification of urban functional zones focuses on the social and economic functions of various urban regions, and divides the overall urban space into a series of regions with various differentiated characteristics, such as commercial, residential, industrial concentration, culture and education, storage and other dominant characteristics. Traditional methods for classifying and identifying urban functional regions are mostly based on remote sensing data, and deep learning is introduced to identify functional regions based on pixel-level classification, object-oriented classification and scene-level classification data. However, remote sensing data only stay on regional natural information. In order to meet the humanistic and social science attributes of urban functional regions, urban function research based on or integrated with social data has become a major trend of current development.

Combining remote sensing image data with spatial information is helpful to identify the urban functional structure and characteristic regions. For example, Aubrecht and Leon Torres [15] introduces a novel top-down approach to classify mixed or residential regions from nighttime light (NTL) images. Using several landscape metrics, Lin et al. [16] extracted land use from Pleiades images to investigate the urban functional landscape pattern in Xiamen, China, and explored the relationships between features of urban functional landscape patterns and population density. Yu and Ng [17] classified land use from Landsat TM images and performed gradient analysis to analyze spatial and temporal urban sprawl dynamics in this city. Yang et al. [18] investigated bag-of-visual-words (BOVW) approaches to land-use classification in high-resolution overhead imagery and proved it a robust alternative that is more effective for certain land-use classes. Zhang et al. [19] applied Linear Dirichlet Mixture Model and Markov process to develop a decomposition model for quantifying mixed semantics of urban scenes through remote sensing data. These studies focus on the spatial features in the city, which only describe the natural characteristics of urban regions, but ignore the effect of human activity. However, in the highly urbanized cities, most land parcels are covered by man-made infrastructures that are a mix between residential, businesses and work function. Such complex urban environments raise a great challenge in understanding urban structure using only remote sensing imagery.

The functional characteristics of a region are not only related to the natural characteristics of the region, but also determined by the characteristics of human activities in the region to a greater extent. However, remote sensing data only describes the natural information characteristics of the geographical region, but cannot reflect the information of human activities. With the rise of Social networks, people have collected a large amount of social data that can reflect the characteristics of human activities, such as POI data, trajectory data, check-ins data and other Social media data. These data used to identify urban functional regions can make up for the deficiency of remote sensing data. Yuan et al., for example, first proposed the framework of Regions of Different Functions (DRoF), using the road network partition method for regional segmentation, and fusing taxi track data and POI data for functional zoning in Beijing [20]. In addition, bus data were also used for further research, and it was found that the method had the best effect when location semantics and motion semantics were considered simultaneously [21]. Gao Song et al. developed a statistical framework, applied LDA topic model, and integrated user check-in on LBSN. Their proposed function based on probabilistic topic modeling revealed the potential structure and location semantics of interest point mixing [22]. Yao et al. [23] combined remote sensing with social media data to propose a novel scene classification framework to identify dominant urban land use type at the level of traffic analysis zone by integrating probabilistic topic models and support vector machine. Similarly, Wei Tu and Hu [24] portrayed urban functional regions by coupling remote sensing imagery and human sensing data to create A new framework by a hierarchical Clustering. Yao Shen et al. [25] describes street network through social media data, constructs urban structural connectivity, and further divides urban functional regions.

POI Data is a very important class of Social Data, and its inherent Category attribute can identify the function of its region to a certain extent. Therefore, POI Data is very suitable for the identification of urban functional regions. Different from the above research, this paper proposes an embedded representation of regional POI based on LDA topic model. Used for urban functional region identification.

## 3    Region Functions Based on LDA Topic Model

This section introduces the relevant concepts and experimental methods of urban functional regions. Firstly, it gives the definition and mathematical expression of keywords in this paper, and then obtains the Block embedding representation method and its mathematical model based on the topic model. Finally, the process of generating functional regions by clustering analysis is explained.

### 3.1    Problem Statement

In this paper, based on POI data and Block partition, LDA model is used as the core method of label, and finally clustering method is used to identify functional regions. First of all, POI data and divided blocks need to be embedded to represent, on the basis of which urban functional regions are defined. Specifically, the relationship between Urban Area (UA), UFR and blocks is shown in Fig. 1:

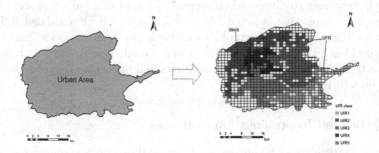

**Fig. 1.** Schematic diagram of the relationships between UA, UFR and block

*Definition 1: Point of Interest.* POI can represent all geographical entities in a city. Let P be the set of all POI data in the study region. Each POI data (p $\in$ P) is defined as $<Cat, Lng, Lat>$, where *Cat* is its category, and if we have m categories, the number of topic words in our model is also m. and respectively represents the longitude and latitude of its location.

*Definition 2: Block.* In this paper, the study region is divided into grid cells of equal size. Each grid cell is called Block and is represented by $b$. The specific position information of a Block can be determined by two boundary points, and the POI data contained in each Block can be replaced by a set. Therefore, Block($b$) is defined as $<Loweleft, Upperright>$ , where *Lowerleft* represents the latitude and longitude at the lower left corner of the Block, and *Upperright* represents the latitude and longitude at the upper right corner of the Block. The specific size and position of the rectangular Block can be determined by these two elements. $b$ is the set of POI data contained in the block, i.e. $b_k = \{P_1, P_2, ..., P_N\}$ Before we merge these similar blocks, we have to realize the embedded representation of blocks and each block can be represented as: $\overline{b_k} = <T_1, T_2, ..., T_n>$.

*Definition 3: UFR.* Urban Functional Region (UFR) aggregates blocks with similar semantics, and the aggregated region is used as a functional region. Each UFR is a collection of the blocks it contains, and blocks in the same UFR represent the similar functions while those in different UFR will represent distinct functions. $UFR: UFR = \{b_1, b_2, ..., b_K\}$.

*Definition 4: Urban Area (UA).* UFR identification problem transforms an urban region into a collection of different functional regions, i.e. a Urban Area: $UA = \{UFR_1, UFR_2, ..., UFR_L\}$.

Therefore, it can be found that, in order to solve the problem of identifying UFR, it is necessary to divide urban regions into grid units of equal size blocks, learn semantic similarity between blocks and aggregate blocks, label different clusters with different functions, obtain multiple functional regions, and finally obtain the identification results of urban functional regions. How to represent blocks so that they can learn their similarities and aggregate them becomes the key to the solution.

## 3.2  Embedded Representation of Blocks

In order to aggregate blocks into several functional regions and obtain functional region identification results, semantic correlation between blocks needs to be learned. The POI data contained in a Block can reflect its semantics, so semantic information between POI data needs to be mined. Therefore, how to correlate POI data with LDA model is the focus of this paper.

In the previous section, we got the partitioning of blocks, and for each Block, we can think of it as a document in the topic model. The information in the Block region is all kinds of POI data. Each POI data (in the pre-processing) is regarded as a word, and POI data has been transformed into vector representation which can be analyzed in probability to obtain the topic vector. i.e. $\overline{t_1} = <word_1, word_2, ..., word_n>$. The practical significance of these topic vectors is the specific function of the city. After we get the topics, we can represent a block by these topics in the form of vector, and each element in the vector represents the proportion of the topic in this block. i.e. $\overline{b_1} = <T_1, T_2, ..., T_n>$. By analyzing the proportion of data of different dimensions in the topic vector, the function distribution under the topic can be obtained, and the keywords can be labeled. The problem of how to confirm the number of functional region types is transformed into determining the number of LDA keywords. Figure 2 show the details. When determining the number of LDA keywords, this paper puts forward the consistency and perplexity degree of the model to evaluate. Consistency refers to the distinction between different topics, which describes the distribution distance between different topics. The higher the consistency, the better. The degree of perplexity means that after text analysis, the trained model analyzes uncertainty of which topics are included in the document. The lower the degree of perplexity, the better.

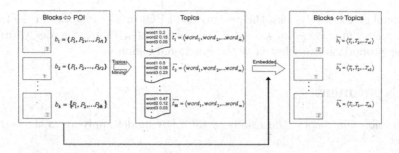

**Fig. 2.** Embedded representation of a block

## 3.3   Functional Region Identification

On the basis of Block embedding representation, the similarity between blocks is analyzed, and the semantically similar blocks are divided into a set, which means that the vector of each Block is clustered. Such a set is our functional region, and finally we realize the visualization and labeling. We choose k-means and spectral clustering algorithms to study the correlation between each Block. The contour coefficients of different K values in the K-means method were analyzed, and suitable K values were selected for clustering operation. Finally, all blocks were divided into K clusters, namely K functional regions. In spectral clustering, besides the number of clusters, the gamma parameter should be determined. For different gamma parameters, the final score CH can be calculated. Both methods prove the feasibility of this model.

In order to achieve the purpose of identifying urban functional regions, after obtaining the clustering results, we also need to associate each cluster with the realistic background, which requires to assign functional labels to the functional regions. We represent region functionality with one or more POI categories with a high proportion of each functional region (cluster). Finally, we label the functional regions.

Set each functional region (clustering) for UFR, which is a collection of blocks it contains. In order to label each UFR, we will first sort the POI number in a UFR and then choose the first j POI categories as follows:

$$N_{pj} = \sum_i p_{ji} \tag{1}$$

Then its proportion is:

$$proj_j = \frac{N_{pj}}{\sum_i N_{pi}} \tag{2}$$

A single functional region means that the proportion of a certain POI type in a functional region unit is more than a threshold T (determined by experiment), and its functional nature is determined by this POI category. Mixed functional region means that the proportion of POI of all categories in the functional region unit does not exceed T, so the first 2–4 POI categories with the highest proportion are selected as the functional region label.

## 4  Experiment

This section will display the results of the LDA model applying to the certain dataset.

### 4.1  Experimental Settings

#### 4.1.1  Data Set

In this paper, POI data of six districts (Baqiao District, Beilin District, Lianhu District, Weiyang District, Xincheng District and Yanta District) in the core of Xi'an City, China, is extracted from Baidu Map. Baidu Map divides these POI categories into two levels, including 19 first-level categories and 127 second-level categories respectively. We use primary classification to analyze regional function. In addition, since the POI data of "natural features" category are very few, and the POI data of "entrance and exit" category have little significance to the function of the description region, we deleted these two types of POI from the data. Finally, our data set contains 17 categories and a total of 141552 POI data. The distribution of POI categories is shown in Fig. 3.

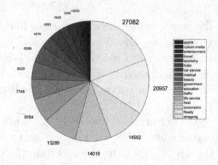

**Fig. 3.** POI category distribution in six districts of Xi'an city

#### 4.1.2  Assessment

In order to verify the accuracy of the recognition results of functional regions and verify the feasibility of the Block embedded representation method, we calculate the evaluation score of the clustering algorithm. The clustering algorithms we choose are K-means algorithm and spectral clustering algorithm, where K is the final cluster number, which has a great influence on the result of functional region identification. Contour coefficient and CH score are used to evaluate, and it is found that under the optimal result, functional region division is similar and the score is also high, which provides a theoretical basis for the robustness of Block embedding representation method. At the same time, this paper found that when K is 6, the clustering result of this experiment is the best.

## 4.2  Experimental Results and Analysis

In this paper, five of the factors that may affect the experimental results are selected for analysis, and five groups of comparative experiments are set up respectively. They are respectively the comparison of different Block sizes with different number of subject words, k-means, comparison of different K values of spectral clustering algorithm, influence of different clustering algorithms on recognition results, and influence of different number of topics in LDA model on results. Finally, we compare the result of functional region identification with baidu map to verify our method.

### 4.2.1  Comparison of Different Block Sizes

In order to obtain functional zoning with more appropriate granularity in urban regions, this paper adopts four blocks of different sizes to conduct experiments, with the sizes of the four blocks being 200 m * 200 m, 500 m * 500 m, 1000 m * 1000 m and 1500 m * 1500 m respectively. K is selected from 2 to 9 for clustering, and the result diagrams of functional region identification are compared and analyzed. Since there is not much difference between the maximum contour coefficients of each size, the actual map of each size Block under the maximum contour coefficients is observed to obtain the most appropriate Block size from both qualitative and quantitative analysis, as shown in Fig. 4.

Figure 4 shows that small Block size leads to finer and more accurate division of functional regions, but too small ones will lead to scattered distribution of functional regions and poor clustering effect. Large Block size and concentrated distribution of each functional region can form a large-scale functional region distribution. However, if the Block size is too large, some functional regions of other categories with small region will be covered and the classification is not fine. In practical applications, we can select blocks of different sizes according to requirements to divide functional regions and balance Block sizes.

For example:

(1) In the figures 500 * 500 and 1000 * 1000, there are no gray blocks in Lianhu District, Xincheng District and Stele Forest District, but the "tourist attraction" functional region at the junction of Xincheng District and Weiyang District is also clearly visible. However, in the figure 1500 * 1500, the gray region occupies a very small region, and the "tourist attraction" functional region in this location disappears.

(2) In the figure of 1500 * 1500, the regions radiating from the three central districts (Lianhu District, Beilin District and Xincheng District) to the periphery can be considered as "shopping-real estate-gastry-life service" functional regions (except one Block in the new city district which is "shopping-corporate-gastry-food" functional region, which is not included). In the

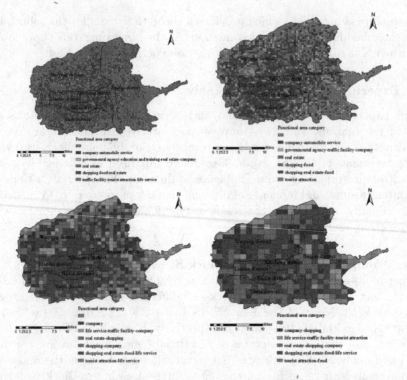

**Fig. 4.** Maximum contour coefficient of four block sizes: 200 m * 200 m, 500 m * 500 m, 1000 m * 1000 m, 1500 m * 1500 m. (Color figure online)

figure of 1000 * 1000 and 500 * 500, there are more scattered "real estate - shopping" functional regions and other functional regions in the central district, while in the figure of 200 * 200, there are more other scattered functional regions and their distribution is dispersed.

It can be found that in the analysis of the actual clustering result graph, the clustering effect is more detailed when the Block size is 500 m * 500 m and 1000 m * 1000 m. Considering the clustering effect under the evaluation of the information provided and contour coefficient, we adopt the Block size of 500 m * 500 m in the subsequent experiment.

### 4.2.2 Feasibility of the Model Under Different Clustering Algorithms

The vector clustering results of each Block of k-means clustering algorithm can effectively quantify the relationship between different blocks, and then obtain the recognition results of functional regions. The evaluation index of contour coefficient is used to evaluate k value, which can be 0.55 under the Block 500 m * 500 m.

In order to analyze the influence of contour coefficients on the results under the K-means clustering algorithm, the contour coefficients were set as small, high and highest, namely, the clustering results when K was 4, 6 and 8 and the block size was 500 m * 500 m, as shown in Fig. 5.

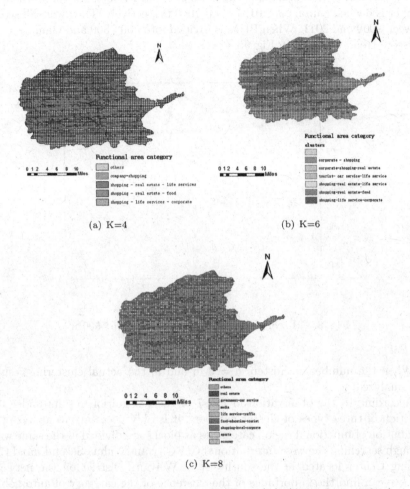

Fig. 5. Clustering results in the case of K = 4/6/8 (Color figure online)

When K is 4, the label of "shopping-life service-company" with industrial attributes is subdivided into the label of "shopping-life service-food". As a result, large red regions appear at the edge of the region in Fig. 4 and 5. However, the red regions are different from the red regions in the city center. At the same time, when k is 8, additional labels with shopping attributes are associated with the company, enterprise and residence, which is very similar to another category of "shopping-life-service-company" label, resulting in redundancy. In contrast, the urban functional zoning with the highest contour coefficient k being 6 is more reasonable in terms of classification.

Similarly, when parameters are traversed in spectral clustering, the gamma value is 0.01, 10 and 1000 respectively, and the number of clusters ranges from 2 to 9. The CH score obtained by analysis is found to be the highest when the number is 6 when gamma is 0.01, 10 and 1000 respectively. They were all scored between 2009 and 2011. When Block is divided into $500 * 500$ and Gamma = 10, CH score trend is shown in Fig. 6:

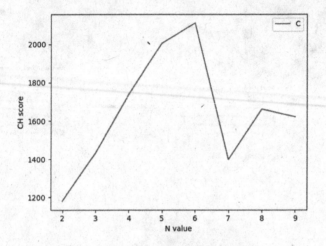

**Fig. 6.** CH score trend at gamma = 10 500 m * 500 m

When the number of clusters n = 3, 6 and 9, the actual clustering results were analyzed:

According to the observation in Fig. 7 of the recognition results under the condition of three types of cluster numbers, it is only necessary to analyze the distribution of functional region categories as blocks are divided in the same way. Through searching the west International Car City and Yuhua Second-hand Car Trading Center located in the southwest of Weiyang District on the network map, we can find the importance of the existence of the category of automobile service-company. However, when the cluster number is 9, the same category of shopping-real estate label appears, and the functional region category appears redundant. Therefore, when the CH index is the highest and there are 6 clusters, the recognition effect of functional regions is better, as shown in Fig. 8:

The k-means clustering result was 6 clusters when the contour coefficient was the highest and the spectral clustering result was 6 clusters when the CH index was the highest, as shown in Fig. 9. The recognition results of functional regions are basically the same under the same number of clusters, and the two clustering algorithms can get very similar results when they tend to the optimal clustering results of Block vector representation set.

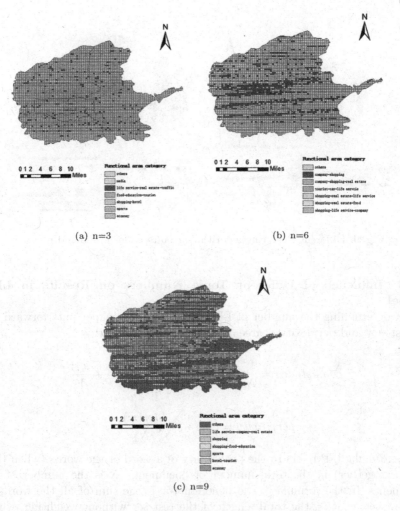

**Fig. 7.** Clustering results under n = 3/6/9

**Fig. 8.** Car service area and second-hand car trading center in southwest Weiyang district

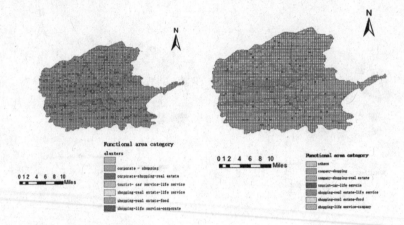

**Fig. 9.** Different clustering algorithms results under k = 6 and n = 6

### 4.2.3 Influence of Different Topic Numbers on Results in LDA Model

When determining the number of LDA keywords, this paper puts forward the consistency and perplexity degree of the model to evaluate:

$$consistency = \frac{2}{N * (N-1)} \sum_{i=2}^{N} \sum_{j=1}^{i-1} log \frac{P(w_i, w_j)}{P(w_j)} \qquad (3)$$

$$perplexity = exp(-\frac{\sum log p(w)}{\sum_d N_d}) \qquad (4)$$

In the formula 3, P refers to the probability of a word or the words (when they appear together) in the total number of documents, N is the number of the documents. In the formula 4, the denominator is the sum of all the words in the test set, that is, the total length of the test set without weighting, where p(w) refers to the probability of each word appearing in the test set. Consistency refers to the distinction between different topics, which describes the distribution distance between different topics. The higher the consistency, the better the outcome. The degree of perplexity means that after text analysis, the trained model analyzes the uncertainty of which topics are included in the document. When the degree of perplexity is lower, the outcome shows better. In this paper, the number of topics was set from 4 to 9 for analysis, and it was found that when the number of topics was 7, consistency and perplexity reached the best level, as shown in Fig. 10.

Next, we will analyse the differences when the topic number differs. The topic number of THE LDA topic model was firstly defined as 4 (at which the consistency and confusion were poor), and the recognition results of functional regions with 6 clusters were analyzed. As shown in Fig. 11, when the number of topics is set to 4, the distribution of the four topics is discrete, and each topic has

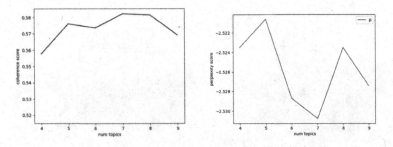

**Fig. 10.** Consistency and perplexity of keywords

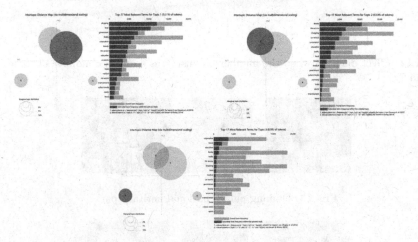

**Fig. 11.** When the number of topics is 4, the representation of topics 1, 2 and 3

its own prominent features. Among them, topic 4 is a special existence, because when the number of topics is 7, there is no topic that takes tourist attractions as the first one in weight ranking. When the number of topics is set as 5, 6, 8, 9 and 10, the analysis of the topic representation also shows that there is no topic represented by tourist attractions. Therefore, a special analysis is made here when the number of topics is 4. In addition, topic 1 shows a strong commercial nature;topic 2 is closely related to life requirement.

The results of functional region identification were compared when the number of topics was 4 and 7. As shown in Fig. 12, when the number of topic is 4, the recognition of tourist attractions is very prominent, while when the number of topic is 7, tourist attractions are clustered into the enterprise-real estate region. However, it is not sensitive enough to identify the region where the car service is located, so it gathers the aforementioned Yuhua second-hand car trading center into the company region.

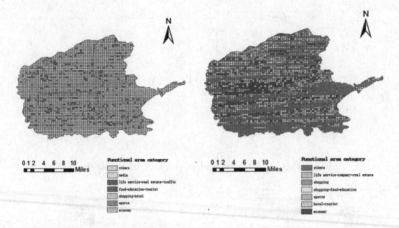

**Fig. 12.** Effect diagram when the number of clusters is 6 and the number of topics is 4 and 7

**Fig. 13.** Daming palace national heritage park

**Fig. 14.** Weiyang palace ruins park

As a key tourism city in China, Xi'an has many tourist attractions. As shown in Fig. 13, daming Palace National Heritage Park is selected for comparison. When the number of topics is 7, it is identified as an enterprise intensive region. However, when the number of topics is 4, it is accurately identified as a tourist spot region. As shown in Fig. 14, weiyang Palace National Heritage Park was selected for comparison. When the number of topics was 7, it was identified as a company or residential region, while when the number of topics was 4, it was accurately identified as a tourist spot region.

Through the above experiments, it is proved that when the number of topics is 7, the effect of functional zoning is the best. In the LDA model, we can find different numbers as the main topics to adjust parameters and analyze the clustering results under the condition of different topic numbers. When some said topic has certain characteristics value, namely the confusion degree is low, the topic has more distinctive characteristics, and the clustering results tend to have high accuracy compared with the actual land use.

## 5     Conclusion

In this paper, we implement a new Block embedded representation method, which combines the POI data of the target city with the LDA model to identify urban region functions. The discovered functional regions contribute to the understanding of complex metropolises and provide lessons for urban planning and facility siting. By using cluster analysis, this paper visualizes the urban functional region, and divides and explains the urban region more conveniently. The clustering results under different topic number and cluster number are analyzed by combining specific landmarks or universally recognized blocks such as the trading center and relic park in southwest Xi'an, which verifies the feasibility of the new method in this paper. Meanwhile, the optimal values of topic number and cluster number under this method are also obtained, which may be helpful for other research work.

In the future research, we can try to improve the user friendliness by upgrading the current offline static mode to the form of dynamic interaction, which can give users more freedom.

**Acknowledgement.** The work is supported by the National Natural Science Foundation of China under Grant No. 61972317, No. 61972318, the Natural Science Foundation of Shaanxi Province of China under Grant No. 2021JM068, the Shaanxi Province Training Program of Innovation and Entrepreneurship for Undergraduates under Grant No. S202110699625.

## References

1. Ye, C., Zhang, F., Mu, L., Gao, Y., Liu, Y.: Urban function recognition by integrating social media and street-level imagery. Environ. Plan. B Urban Anal. City Sci. **48**(6), 1430–1444 (2021). https://doi.org/10.1177/2399808320935467

2. Ramaswami, A., Russell, A., Culligan, P., Rahul Sharma, K., Kumar, E.: Meta-principles for developing smart, sustainable, and healthy cities. Science **352**(6288), 940–943 (2016). https://doi.org/10.1126/science.aaf7160. Funding Information: The authors are grateful for support from NSF (Partnership for International Research and Education award 1243535 and Sustainability Research Networks award 1444745) and from the U.S. Agency for International Development and the National Academy of Sciences (Partnership for Enhanced Engagement in Research subgrant 2000002841)

3. Hu, S., et al.: Urban function classification at road segment level using taxi trajectory data: a graph convolutional neural network approach. Comput. Environ. Urban Syst. **87**, 101619 (2021). https://doi.org/10.1016/j.compenvurbsys.2021. 101619

4. Han, J., Chen, W.-Q., Zhang, L., Liu, G.: Uncovering the spatiotemporal dynamics of urban infrastructure development: a high spatial resolution material stock and ow analysis. Environ. Sci. Technol. **52**(21), 12122–12132 (2018). https://doi.org/10.1021/acs.est.8b03111

5. Xu, G., Zhou, Z., Jiao, L., Zhao, R.: Compact urban form and expansion pattern slow down the decline in urban densities: a global perspective. Land Use Policy **94**, 104563 (2020). https://doi.org/10.1016/j.landusepol.2020.104563

6. Xu, Y., Olmos, L., Abbar, S., Gonzalez, M.C.: Deconstructing laws of accessibility and facility distribution in cities. Sci. Adv. **6** (2020). https://doi.org/10.1126/sciadv.abb4112

7. La Rosa, D., Privitera, R.: Characterization of non-urbanized areas for land-use planning of agricultural and green infrastructure in urban contexts. Landscape Urban Plann. **109**, 94–106 (2013). https://doi.org/10.1016/j.landurbplan.2012.05.012

8. Henderson, J., Venables, A., Regan, T., Samsonov, I.: Building functional cities. Science **352**, 946947 (2016)

9. Morawska, L., et al.: Towards Urbanome the genome of the city to enhance the form and function of future cities. Nature Commun. **10**, 1–3 (2019). https://doi.org/10.1038/s41467-019-11972-6

10. Habitat, U.: Urbanization and development: emerging futures. World Cities Report 2016 (2016)

11. Ziwei, G., Weiwei, S., Penggen, C., Gang, Y., Xiangchao, M.: Identify urban functional zones using multi feature latent semantic fused information of high-spatial resolution remote sensing image and poi data. Remote Sens. Technol. Appl. **36**(3), 618 (2021). https://doi.org/10.11873/j.issn.1004-0323.2021.3.0618

12. Zhang, D., et al.: Identifying region-wide functions using urban taxicab trajectories. ACM Trans. Embed. Comput. Syst. **15**, 1–19 (2016). https://doi.org/10.1145/2821507

13. Volgmann, K., Rusche, K.: The geography of borrowing size: exploring spatial distributions for German urban regions. Tijdschrift voor Economische en Sociale Geografie **111**, 60–79 (2019). https://doi.org/10.1111/tesg.12362

14. Jiang, S., Alves, A., Rodrigues, F., Ferreira, J., Pereira, F.C.: Mining point-of-interest data from social networks for urban land use classification and disaggregation. Comput. Environ. Urban Syst. **53**, 36–46 (2015). https://doi.org/10.1016/j.compenvurbsys.2014.12.001. Special Issue on Volunteered Geographic Information

15. Aubrecht, C., León Torres, J.A.: Evaluating multi-sensor nighttime earth observation data for identification of mixed vs. residential use in urban areas. Remote Sens. **8**, 114 (2016). https://doi.org/10.3390/rs8020114

16. Lin, T., et al.: Spatial pattern of urban functional landscapes along an urban-rural gradient: a case study in Xiamen City, China. Int. J. Appl. Earth Obs. Geoinf. **46**, 22–30 (2016). https://doi.org/10.1016/j.jag.2015.11.014

17. Yu, X., Ng, C.: Spatial and temporal dynamics of urban sprawl along two urban-rural transects: a case study of Guangzhou, china. Land-scape Urban Plann. **79**, 96–109 (2007). https://doi.org/10.1016/j.landurbplan.2006.03.008

18. Yang, Y., Newsam, S.: Bag-of-visual-words and spatial extensions for land-use classification, pp. 270–279 (2010). https://doi.org/10.1145/1869790.1869829

19. Zhang, X., Du, S.: A linear Dirichlet mixture model for decomposing scenes: application to analyzing urban functional zonings. Remote Sens. Environ. **169**, 37–49 (2015). https://doi.org/10.1016/j.rse.2015.07.017

20. Yuan, J., Zheng, Y., Xie, X.: Discovering regions of different functions in a city using human mobility and POIs. In: Proceedings of the 18th ACM SIGKDD International Conference on Knowledge Discovery and Data Mining, KDD 2012, pp. 186–194. Association for Computing Machinery, New York (2012). https://doi.org/10.1145/2339530.2339561

21. Yuan, N., Zheng, Y., Xie, X., Wang, Y., Zheng, K., Xiong, H.: Discovering urban functional zones using latent activity trajectories. IEEE Trans. Knowl. Data Eng. **27**, 712–725 (2015). https://doi.org/10.1109/TKDE.2014.2345405

22. Gao, S., Janowicz, K., Couclelis, H., et al.: Extracting urban functional regions from points of interest and human activities on location-based social networks. Trans. GIS **21**, 446–467 (2017). https://doi.org/10.1111/tgis.12289

23. Yao, Y., et al.: Sensing spatial distribution of urban land use by integrating points-of-interest and Google Word2Vec model. Int. J. Geograph. Inf. Sci. **31**, 1–24 (2016). https://doi.org/10.1080/13658816.2016.1244608

24. Tu, W., et al.: Portraying urban functional zones by coupling remote sensing imagery and human sensing data. Remote Sens. **10**, 141 (2018). https://doi.org/10.3390/rs10010141

25. Shen, Y., Karimi, K.: Urban function connectivity: characterisation of functional urban streets with social media check-in data. Cities **55**, 9–21 (2016). https://doi.org/10.1016/j.cities.2016.03.013

# A Data-to-Text Generation Model
# with Deduplicated Content Planning

Mengda Wang[1], Jianjun Cao[2(✉)], Xu Yu[2], and Zibo Nie[2]

[1] School of Computer Science, Nanjing University of Information Science and Technology, Nanjing, China
[2] The Sixty-Third Research Institute, National University of Defense Technology, Nanjing, China
caojj@nudt.edu.cn

**Abstract.** Texts generated in data-to-text generation tasks often have repetitive parts. In order to get higher quality generated texts, we choose a data-to-text generation model with content planning, and add coverage mechanisms to both the content planning and text generation stages. In the content planning stage, a coverage mechanism is introduced to remove duplicate content templates, so as to remove sentences with the same semantics in the generated texts. In the text generation stage, the coverage mechanism is added to remove the repeated words in the texts. In addition, in order to embed the positional association information in the data into the word vectors, we also add positional encoding to the word embedding. Then the word vectors are fed to the pointer network to generate content template. Finally, the content template is inputted into the text generator to generate the descriptive texts. Through experiments, the accuracy of the content planning and the BLEU of the generated texts have been improved, which verifies the effectiveness of our proposed data-to-text generation model.

**Keywords:** Data-to-text generation · Content planning · Coverage mechanism · Positional encoding

## 1 Introduction

Data-to-text generation (D2T) is one of the three major research directions in Natural Language Generation (NLG), and the other two directions are text-to-text generation and graph-to-text generation. Through entering structured data, such as a table, the model generates a fluent and accurate descriptive texts [1]. In the era of information explosion, D2T provides great convenience for data mining.

Data-to-text generation methods are generally divided into traditional template-based methods and end-to-end neural models. Traditional methods select and sort structured data through content planning, so the generated texts are well organized and have high accuracy, but with poor fluency. On the contrary, the data-to-text generation models

The National Natural Science Foundation of China (61371196), National Science and Technology Major Project (2015ZX01040-201).

based on neural networks do not interfere with the data, and directly allow the data-driven model to generate text. The generated texts usually have a certain fluency, but the texts which generated by the neural network do not describe the input data accurately. After [2] combines traditional methods and neural network models, the idea of incorporating content planning into neural models is proposed. This new idea divides the entire end-to-end data-to-text generation model into two stages: content planning and text generation. This idea solves the respective shortcomings of the previous two types of models. Data-to-text generation methods that incorporate content planning have received increasing attention from researchers.

With the advent of recurrent neural networks and sequence-to-sequence models, the processing ability of neural networks for text has been significantly improved. The memory characteristic of recurrent neural network can easily capture the semantic relationships between the preceding and following tokens. The sequence-to-sequence model makes it possible to process text of variable lengths. Recurrent neural networks combined with sequence-to-sequence models and integrated attention mechanisms have become a hot point some time ago. However, the shortcomings of recurrent neural networks and sequence-to-sequence framework also become inherent shortcomings of these models. Recurrent neural networks have limited memory capabilities, making it difficult to remember all semantic relationships among texts. The sequence-to-sequence framework also suffers from the Out-Of-Vocabulary (OOV) problem which mentioned in [3], and attention mechanism is easy to generate repeated texts. What's more, the semantic relationship between the front and rear tokens in structured data is not very obvious, but there are some logical associations in the positional relationship. Therefore, the recent D2T with content planning methods still have space for improvement.

In view of the above problems, we add the positional encoding from Transformer to our model to embed data which have fixed structure. Positional encoding will better encode the positional information of tokens into word vectors, and select important tokens through self-attention mechanism. Due to the shortcomings of attention mechanism, it is easy to generate repetitive content. In order to avoid generating repeated content template and descriptive texts, we add coverage mechanism to the sequence to sequence which in the two stages of content planning and text generation. In the text generation stage, we choose long-short-term memory (LSTM) network with better memory characteristics to capture the semantic relationship in the content template for increasing the accuracy and fluency of the generated texts.

## 2   Related Work

A few years ago, D2T focused on researching holistic end-to-end neural network models. Most researchers believed that even without the use of content selection and content planning, only neural language models or machine translation can generate fluent and on-target text [4–7].

Since [1] proposed the RotoWire dataset and challenged D2T experiments on this dataset in 2017. Although the texts generated by the current data-to-text generation neural model was fluent and readable, the content described by the generated text was different from the content pointed by the data. Therefore, researchers began to focus on how to improve the content quality of the generated texts. Considering that the texts generated by traditional methods have good performance in content quality, content planning is gradually integrated into neural networks.

[2] proposed a Neural Content Planning (NCP) model that added content selection and content planning modules to enhance the effect of the model, which split the data-to-text generation neural model into two stages of content planning and text generation. This method made the model more efficient and enhanced the organization of the generated texts structure. The model used a sequence-to-sequence model as a framework in both content planning and text generation stages, and an attention mechanism was used to select data. In addition, an attention mechanism and a copy mechanism were added in the text generation stage to ensure the fluency of the generated texts. On the basis of [2, 8] proposed a numerical data understanding and important information verification module. When model encoded the data, a numerical representation of the context was designed to enhance the understanding of the entity, and an effective reward was designed to validate the content planning results. Module solved the problem that NCP lacked the understanding of the numerical value in the context and the lack of targeted optimization objectives for content planning. [9] proposed a data-to-text generation neural model with dynamic programming which can dynamically generate content templates from data, by adding a dynamic content planning mechanism and a reconstruction mechanism to the content planning module and text generation module of the NCP model respectively. During text generation, the model reconstructed the order of all entities to increase the fluency of the generated texts.

Data-to-text generation models with content planning show good performance in data-to-text generation tasks. Because these models first fit unordered structured data to a content template, and again fit to the target texts, and two fittings greatly enhance the text generation capabilities of these models.

With the advent of more and more deep learning methods, the effectiveness of data-to-text generative neural models had been improved. [10] used Transformer to generate content template, and used the self-attention mechanism to focus on the order of data which need to be expressed to generate content template. This model enhanced the content planning ability of the model, and further improved the accuracy of the generated texts. In order to solve the problem of repeated text generation, [11] introduced a coverage mechanism in the text generation stage to track the historical attention weights, that paid attention to the entire content template as generating texts. Since the calculation of the attention mechanism at different time steps is independent without considering the historical attention weights distribution, the sequence-to-sequence models with attention mechanism always have the problem mentioned in [11], which result that the generated texts contain part of the information of the input data or repeated content. We referred to the coverage mechanism which applied to machine translation tasks in the [12], and proposed a new coverage mechanism applied to the content planning stage. The coverage method reduced the attention weights of some input data, and focused on unused information as much as possible during the content planning phase.

## 3   Proposed Model

We will explain each stage of our model in detail. Our model is based on NCP model [2]. The framework of our proposed model is shown in Fig. 1.

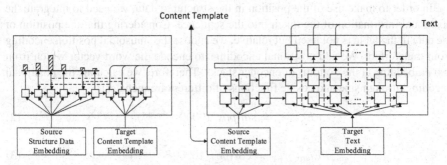

**Fig. 1.** An overview of our model

The whole model is divided into three parts. (1) Word embedding. Extract latent semantics from input structured data through text embedding and positional encoding. (2) Content planning. The coverage mechanism integrated into the pointer network accurately fits the target content template to complete the content planning. (3) Text generation. The text generator is composed of a Bi-LSTM network with copy mechanism and coverage mechanism. The content template is fed to the text generator for generating accurate and fluent descriptive texts.

### 3.1   Word Embedding

The input to our model is a table of unordered records, each record $r_i$ has four features including its TYPE, ENTITY, VALUE and H/V, represented as $\{r_{j,k}\}_{k=1}^{4}$. Following previous work in [2], we embed features into vectors by using word embedding and positional encoding.

Let $r = \{r_j\}_{j=1}^{|r|}$ represent a table of input records. We use a multilayer perceptron to obtain a vector representation $r_i$ for each record:

$$r_j = \text{ReLU}(W_r[r_{j,1}; r_{j,2}; r_{j,3}; r_{j,4}] + b_r) \tag{1}$$

where $[r_{j,1}; r_{j,2}; r_{j,3}; r_{j,4}]$ represents vector representations to be concatenated a vector, $W_r \in \mathbb{R}^{n \times 4n}$, $b_r \in \mathbb{R}^n$ are parameters, and ReLU is the activation function.

Distiawan B et al. [11] found that using RNN for word embedding is not an effective solution. Because structured data is unordered and there is no clear semantic relationship between the previous token and the next token. However, structured data has a certain relationship in position, so we use text embedding and position encoding from [13] to embedding on structured data.

In order to make use of the position in the structured data, we need to integrate the positional information of the token into the sequence. Considering that the position of the data in the table is not fixed but relative, we choose the sinusoidal position encoding from transformer. We add positional encodings to encode the word vector which from text embedding at the bottoms of encoder stacks. Therefore, we calculate the positional encoding through sine and cosine functions of different frequencies:

$$PE_{(pos,2i)} = \sin(pos/10000^{2i/d_{model}}) \tag{2}$$

$$PE_{(pos,2i+1)} = \cos(pos/10000^{2i/d_{model}}) \tag{3}$$

where $pos$ and $i$ are position and dimensions, respectively.

In order to better extract the dependencies between records, we introduce a content selection gate that uses attention mechanism to select the content required for content planning in [2].

$$\alpha_{j,k} \propto \exp(r_j^T W_a r_k) \tag{4}$$

$$c_j = \sum_{k \neq j} \alpha_{j,k} r_k \tag{5}$$

$$r_j^{att} = W_g[r_j; c_j] \tag{6}$$

where $W_a \in \mathbb{R}^{n \times n}$, $W_g \in \mathbb{R}^{n \times 2n}$ are parameter matrices, and $\sum_{k \neq j} \alpha_{j,k} = 1$.

Next, we use the content selection gate mechanism for $r_j$, and obtain a new record representation $r_j^{cs}$ by via:

$$g_j = sigmoid(r_j^{att}) \tag{7}$$

$$r_j^{cs} = g_j \odot r_j \tag{8}$$

where $g_j \in [0, 1]^n$ limits the amount of information output from $r_j$. The framework of word embedding is shown in Fig. 2.

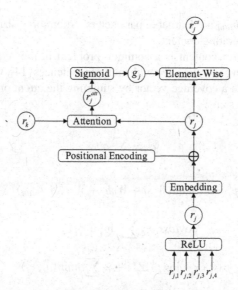

**Fig. 2.** Word embedding

## 3.2 Content Planning

Our model learns such content templates from training data. The framework of content planning is shown in Fig. 3.

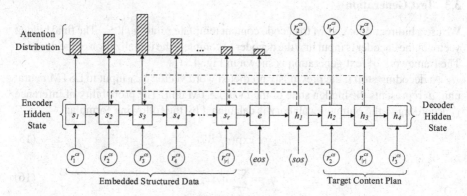

**Fig. 3.** Content planning

The content template consists of a series of pointers $z$. $z = z_1, z_2, \ldots, z_{|z|}$ represents the content template sequence. Each $z_k$ points to an input record, i.e., $z_k \in \{r_j\}_{j=1}^{|r|}$.

$$e_i^t = v^T \tanh(W_s s_i + W_h h_t + b_{attn}) \tag{9}$$

$$a^t = \text{softmax}(e^t) \tag{10}$$

where $v$, $W_s$, $W_h$ and $b_{attn}$ are learnable parameters. Attention distribution plays a role for the decoder in generating content.

Generating repetitive content is a common problem in the sequence-to-sequence model, especially when generating text with multiple sentences [14]. We use the coverage mechanism to generate a coverage vector by summing the attention distribution of all previous decoder time steps:

$$c^t = \sum_{j}^{t-1} a^j \tag{11}$$

$$p_i^t = v^T \tanh(W_s s_i + W_h h_t + W_c c_i^t + b_{attn}) \tag{12}$$

$$covloss_t = \sum_i \min(a_i^t, c_i^t) \tag{13}$$

$$loss_t = -\log p(W_t^*) + \lambda \sum_i \min(a_i^t, c_i^t) \tag{14}$$

where $c^t$ is the unnormalized distribution of descriptive text in the source dataset, which indicates the degree of coverage of these words received from the attention mechanism so far, and $W_c$ is a parameter vector with the same length as $v$. This ensures that the current decisions of the attention mechanism are informed by the reminder of their previous decisions, which are summarized in the $c^t$.

## 3.3  Text Generation

We use a bidirectional LSTM to encode content template $z$ into $\{e_k\}_{k=1}^{|z|}$. The final hidden vector of the encoder is input into the decoder to complete the initialization of the decoder. The framework of text generation is shown in Fig. 4.

At decoding stop $t$, the previously predicted word vector $y_{t-1}$ input to LSTM neural unit. $d_t$ represents the hidden state of the $t$ - th LSTM unit. The probability of inference $y_t$ according to the output vocabulary is calculated by the following formula:

$$\beta_{t,k} \propto \exp(d_t^T W_b e_k) \tag{15}$$

$$q_t = \sum_k \beta_{t,k} e_k \tag{16}$$

$$d_t^{att} = \tanh(W_d[d_t; q_t] \tag{17}$$

$$p_{gen}(y_t|y_{<t}, z, r) = \text{softmax}_{y_t}(W_y d_t^{att} + b_y) \tag{18}$$

where $\sum_k \beta_{t,k}=1$, $W_b \in \mathbb{R}^{n \times n}, W_d \in \mathbb{R}^{n \times 2n}, W_y \in \mathbb{R}^{n \times |v_y|}, b_y \in \mathbb{R}^{|v_y|}$ are parameters, and $|v_y|$ is the output vocabulary size.

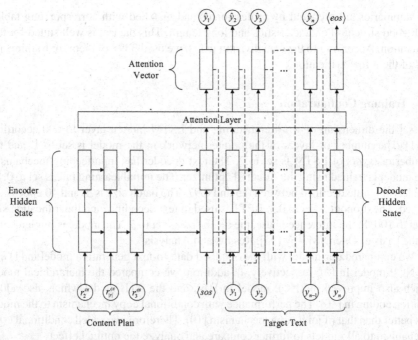

**Fig. 4.** Text generation

### 3.4 Training and Inference

The goal of our model training is to maximize the log Likelihood of the input table record $r$ generating content template and the input content template and form record generating text:

$$\max \sum_{(r, z, y) \in \mathbf{D}} \log p(z|r) + \log p(y|r, z) \tag{19}$$

where $D$ is training examples.

During inference, the generated text are predicted by the following formula

$$\hat{z} = \arg\max_{z'} p(z'|r) \tag{20}$$

$$\hat{y} = \arg\max_{y'} p(y'|r, \hat{z}) \tag{21}$$

where $z'$ represents content template, and $y'$ represents generated texts candidates. In addition, we add beam search mechanism to get the best results.

## 4 Experiment

### 4.1 Dataset

We used the RotoWire dataset to train and evaluate our model. The RotoWire dataset collects many basketball game data. The dataset contains tables and summaries, of which the summaries are prepared by professionals and matched with corresponding tables, with good structural characteristics and long length. This dataset is well suited for text generation. According to the data division in [1], we used 3398 of them for training and 727 of them for verification.

### 4.2 Training Configuration

We set the dimension of word embedding and LSTM hidden layer to 600 according to [1]. The number of layers of the pointer network in the model is set to 1, and the number of layers of LSTM is set to 2. The text decoder has Inputfeeding mechanism. The model is trained using the Adagrad optimizer. The initial learning rate is set to 0.15. The learning rate declined between 0.5 and 0.97. The batch size is 5, and 40 epochs are trained. The dropout is set to 0.3. BPTT is used in text decoding, and the truncation size is set to 100. In the inference phase, the beam size is set to 5. The model is implemented on the basis of OpenNMT-py [15]. Results and Analysis.

We compared our model with the Seq2Seq data-to-text generation model in [1] and the NCP model in [2], respectively. In addition, we compared the hierarchical model which also improved the NCP model in [16], and the ATL model which also added position coding in [10]. The performance with conditional copy mechanism to the model was better than that of joint copy mechanism [10]. Therefore, we added conditional copy mechanism to all models to further compare and analyze the optimal effect.

We use accuracy and perplexity (PPL) to evaluate the training of the model for verifying the effectiveness of the model, which in Table 1.

**Table 1.** Accuracy and perplexity (PPL) in training

| Model | Content planning | | Text generation | |
|---|---|---|---|---|
| | Accuracy | PPL | Accuracy | PPL |
| NCP | 72.8405% | 2.8948 | 58.6674% | 7.5726 |
| Our model (positional) | 73.3973% | 2.7516 | 58.6737% | 7.3065 |
| Our model (coverage) | 72.8486% | 2.8951 | 58.8540% | 7.3412 |
| Our model (Positional + Coverage) | 74.3825% | 2.6861 | 58.9313% | 7.5160 |

After adding positional encoding to word embedding, the output accuracy of content planning is increased from 72.8405% to 73.3973%, and the perplexity is reduced from 2.8948 to 2.7516. Because after adding the positional encoding, the semantic information contained in the position association relationship is fixed and extracted. The decoder in the content planning stage can generate a content plan according to more information. When only the coverage mechanism is added, the accuracy rate in the text generation stage increases from 58.6674% to 58.8540%, and the confusion degree decreases from 7.5726 to 7.3412. The coverage mechanism removes the duplicate content in the generated text which make the text more fluent. When the positional encoding and coverage mechanisms are added at the same time, the accuracy is increased from 73.3973% to

74.3825%, and the confusion is reduced from 2.7516 to 2.6861. The coverage mechanism added in the content planning removes the repetitively generated sentences. However, the generated text does not contain a lot of repeated content, so the improvement effect is not obvious.

In order to evaluate the coherence of the text generated by our model, we evaluated the CS (Content Selection), RG (Relation Generation), CO (Content Ordering) [1] and BLEU [17] in the valid set. The results are shown in Table 2, and part of texts generated by the model are shown in Table 3.

**Table 2.** Evaluation results by model on CS, RG, CO and BLEU metrics

| Model | RG | | CS | | CO | BLEU-4 |
|---|---|---|---|---|---|---|
| | # | P% | P% | R% | D% | |
| WS-2017 | 24.0 | 75.1 | 28.1 | 35.9 | 15.3 | 0.141 |
| NCP | 34.1 | 87.2 | 32 | 47.3 | 17.2 | 0.165 |
| Hierarchical | 21.2 | **89.1** | **38.7** | **51.5** | 18.7 | **0.173** |
| ATL | 34.2 | 87.5 | 34.1 | 50.2 | **20.6** | 0.161 |
| Our model (Positional + Coverage) | **34.6** | 87.8 | 33.6 | 49.6 | 18.3 | 0.168 |

**Table 3.** Generated texts and original texts

| The Generated Texts | The Original texts |
|---|---|
| The Toronto Raptors defeated the Philadelphia 76ers, 122–95, at Air Canada Centre on Friday. The Raptors (11–6) checked in to Saturday's contest with only two road victories in 14 tries away from the Air Canada Centre, while the Raptors (4–14) checked in to Friday with only five road victories of the season. The 76ers (4–14) had come in to Tuesday's contest having lost five of their last seven games … | The host Toronto Raptors defeated the Philadelphia 76ers, 122–95, at Air Canada Center on Monday. The Raptors came into this game as a monster favorite and they did n't leave any doubt with this result. Toronto just continuously piled it on, as they won each quarter by at least four points. The Raptors were lights - out shooting, as they went 55% from the field and 68% from three - point range … |

Table 2 shows that our model has a slight improvement in each indicator after adding positional encoding and coverage mechanism respectively. RG evaluates the authenticity of the generated text, CS evaluates the performance of content selection, and co evaluates the quality of content templates, and the BLEU evaluates the fluency of the text, and the effect is not obvious after adding positional encoding, because the positional encoding only has an effect on the extraction of semantic information in the data. The decoder in the content planning stage can generate a content plan according to more information. After continuing to add the coverage mechanism, the effect is significantly improved than the position encoding, because the coverage mechanism removes the repetitive content in the generated text, making the text more fluent. However, since the generated text itself contains less repetitive content, the improvement effect is not very obvious. Compared with the ATL model, all indicators except CO have slightly improved. This means that location coding has played a full role in the model, and the addition of coverage mechanism makes the generated text remove duplicate characters, which improves the effect of the model. Compared with the hierarchical model, our model only has advantages in Co. Through our analysis, this is because the hierarchical model uses auxiliary tasks to improve the NCP model, and the auxiliary tasks have a good response in data to text generation. However, excessive use of auxiliary tasks will make the generated text unreal.

Table 3 shows the generated text and the reference text in the dataset. The generated text contains certain information and also has a certain degree of fluency between sentences. However, compared with the reference text, there is still a certain distance.

## 5    Conclusion

We presented a data-to-text generation model which enhances content planning modules by adding positional encoding and coverage mechanism. Experimental results based on metrics demonstrate that our model have better generation ability in terms of the number of relevant facts contained in the output texts. Since content planning with coverage mechanism can concise generated texts effetely.

In the experiment, we found that our model has a significant improvement in content planning. However, the text generation stage reduces the overall performance. A good text generator will make the quality of the generated texts improved. The pretrained language models have an obvious effect improvement in area of text generation. In the future, in order to produce texts with higher quality, we will spend more effort using pre-trained language model in the text generation stage.

## References

1. Wiseman, S., Shieber, S.M., Rush, A.M.: Challenges in data-to-document generation. In: Proceedings of the 2017 Conference on Empirical Methods in Natural Language Processing, pp. 2253–2263 (2017)
2. Puduppully, R., Dong, L., Lapata, M.: Data-to-text generation with content selection and planning. In: Proceedings of the 33rd AAAI Conference on Artificial Intelligence, pp. 6908–6915 (2019)
3. Peng, L., Liu, Q., Lv, L., Deng, W., Wang, C.: An abstractive summarization method based on global gated dual encoder. In: Zhu, X., Zhang, M., Hong, Y., He, R. (eds.) NLPCC 2020. LNCS (LNAI), vol. 12431, pp. 355–365. Springer, Cham (2020). https://doi.org/10.1007/978-3-030-60457-8_29
4. Bengio, Y., Ducharme, R., Vincent, P., et al.: A neural probabilistic language model. J. Mach. Learn. Res. 3, 1137–1155 (2003)
5. Mikolov, T., Karafiát, M., Burget, L., et al.: Recurrent neural network based language model. In: Proceedings of the 11th Annual Conference of the International Speech Communication Association, pp. 1045–1048 (2010)

6.  Sutskever, I., Martens, J., Hinton, G.: Generating text with recurrent neural networks. In: Proceedings of the 28th International Conference on International Conference on Machine Learning, pp. 1017–1024 (2011)

7.  Lebret, R., Grangier, D., Auli, M.: Neural text generation from structured data with application to the biography domain. In: Proceedings of the 2016 Conference on Empirical Methods in Natural Language Processing, pp. 1203–1213 (2016)

8.  Gong, H., Wei, B., Xiaocheng, F., et al.: Enhancing content planning for table-to-text generation with data understanding and verification. In: Proceedings of the 2020 Conference on Empirical Methods in Natural Language Processing, pp. 2905–2914 (2020)

9.  Chen, K., Li, F., Hu, B., et al.: Neural data-to-text generation with dynamic content planning. Knowl. Based Syst. **215**, 106610 (2021)

10. Xiaohong, X., Ting, H., Huazhen, W., et al.: Research on data-to-text generation based on transformer model and deep neural network. J. Chongqing Univ. **43**(07), 91–100 (2020)

11. Distiawan, B.T., Jianzhong, Q., Rui, Z.: Sentence generation for entity description with content-plan attention. In: Proceedings of the 34th AAAI Conference on Artificial Intelligence, pp. 9057–9064 (2020)

12. Zhaopeng, T., Zhengdong, L., Yang, L., et al.: Modeling coverage for neural machine translation. In: Proceedings of the 54th Annual Meeting of the Association for Computational Linguistics, pp. 76–85 (2016)

13. Vaswani, A., Shazeer, N., Parmar, N., et al.: Attention is all you need. In: Advances in Neural Information Processing Systems, pp. 5998–6008 (2017)

14. Abigail, S., Peter, J.L., Christopher, D.M.: Get to the point: summarization with pointer-generator networks. In: Proceedings of the 55th Annual Meeting of the Association for Computational Linguistics, pp. 1073–1083 (2017)

15. Klein, G., Kim, Y., Yuntian, D., et al.: OpenNMT: open-source toolkit for neural machine translation. In: Proceedings of the 55th Annual Meeting of the Association for Computational Linguistics-System Demonstrations, pp. 67–72 (2017)

16. Rebuffel, C., Soulier, L., Scoutheeten, G., Gallinari, P.: A hierarchical model for data-to-text generation. In: Jose, J.M., et al. (eds.) ECIR 2020. LNCS, vol. 12035, pp. 65–80. Springer, Cham (2020). https://doi.org/10.1007/978-3-030-45439-5_5

17. Papineni, K., Roukos, S., Ward, T., et al.: BLEU: a method for automatic evaluation of machine translation. In: Proceedings of the 40th Annual Meeting of the Association for Computational Linguistics, Philadelphia, pp. 311–318 (2002)

# Clustering-Enhanced Knowledge Graph Embedding

Fuwei Zhang[1,2], Zhao Zhang[3], Fuzhen Zhuang[4], Jingjing Gu[5], Zhiping Shi[6], and Qing He[1,2(✉)]

[1] Key Lab of Intelligent Information Processing of Chinese Academy of Sciences (CAS), Institute of Computing Technology, Chinese Academy of Sciences, Beijing 100190, China
{zhangfuwei20g,heqing}@ict.ac.cn
[2] University of Chinese Academy of Sciences, Beijing 100049, China
[3] Institute of Computing Technology, Chinese Academy of Sciences, Beijing 100190, China
zhangzhao2021@ict.ac.cn
[4] Xiamen Institute of Data Intelligence, Xiamen, China
zhuangfuzhen@buaa.edu.cn
[5] Nanjing University of Aeronautics and Astronautics, Nanjing, China
gujingjing@nuaa.edu.cn
[6] Capital Normal University, Beijing 100048, China

**Abstract.** Knowledge graph embedding (KGE) is a task to transform the symbolic entities and relations in Knowledge Graphs (KGs) into low-dimensional vectors, which facilitates the use of KGs in downstream applications. However, most existing models ignore the semantic correlations among similar entities and relations. Indeed, we find there exist semantically similarities in entities or relations. To take advantage of these semantic correlations, we utilize the entity cluster and relation cluster information to enhance traditional KGE models. Particularly, we first use an unsupervised clustering algorithm to get the entity and relation clusters. Then we decompose the representation of the entities or relations into two parts, one is from the entities or relations themselves, and the other is from the cluster information. By this way, the entities or relations can make full use of both the common information of their similar entities or relations and the unique information from themselves. Extensive experiments on popular benchmarks demonstrate the effectiveness of the proposed models.

**Keywords:** Knowledge graph · Clustering · Knowledge graph embedding · Knowledge representation

## 1 Introduction

Recently, knowledge graphs (KGs) are widely used in many domains of downstream applications, e.g., information extraction [1,2], relation extraction [3–6],

T. Li et al. (Eds.): BigData 2022, CCIS 1709, pp. 104–123, 2022.
https://doi.org/10.1007/978-981-19-8331-3_7

recommender systems [7–10], question answering [2,11] and cross-lingual sentiment analysis [12]. The KGs are multi-relational graphs that are make up of entities and relations. And the entities and relations are the nodes and edges in the multi-relational graphs, respectively. Usually, a fact in KG is represented as a triplet $(h, r, t)$. And the facts are extracted from the real world. For instance, (*Beijing, partOf, China* represents the entity *Beijing*) is a part of the entity *China* with respect to the relation *partOf*.

Some modern KGs, such as Freebase [13], WordNet [14], DBpedia [15], YAGO [16] and Google's KGs[1], are available. These data sets are large-scale ones. Some of them have even more than one trillion triplets. Although there is a huge number of triplets, the KGs are still incomplete. To solve the incompleteness problem, people have been motivated to work on Knowledge Graph Completion (KGC). Thus knowledge Graph Embedding (KGE) is a sub-task of KGC, which transforms the symbolic entities and relations to low-dimensional embeddings to facilitate the computation on machines.

In the past, there have been a number of studies on the topic of KGE, such as TransE [17], TransH [18], RotatE [19], DistMult [20] and ComplEx [21], etc. Most of them directly use the embeddings without the related information among similar entities or relations, e.g., the clustering information of entities or relations. Along this line, Zhang et al. [22] used the rich information of the relation embeddings and put forward a Hierarchical Relation Structure (HRS) in 2018. Our work is to concentrate on the semantic information in both entities and relations. Specifically, we try to learn the embeddings with the clustering information of entity clusters and relation clusters. With the information of entity clusters and relation clusters, there are three variants of our models: 1) use Shared Entity Cluster (SEC) embeddings; 2) use Shared Relation Cluster (SRC) embeddings; 3) use Shared Entity & Relation Cluster(SERC) embeddings. Here are two parts of our embeddings:

1. **Entity and Relation Embeddings**
   Entity/Relation Embeddings are low-dimension vectors, which represent the symbolic entities and relations. These embeddings are widely used in nearly all the KGE models.
2. **Shared Entity/Relation Cluster Embeddings**
   Shared Entity/Relation Cluster Embeddings utilize the latent related information of similar entities or relations. We use clustering algorithm (e.g., k-means) to obtain them. The entities or relations in the same cluster share the same embedding.

Figure 1 shows an example of some clustering information in a fact triple sampled from Freebase [13]. We can see that there are similar entities and relations in a KG. If we use latent information properly, i.e., using a shared embedding to represent similar entities, we may enhance the KGE task. To further investigate the influence of clustering information in similar entities and relations, we design some experiments. In our experiments, we use TransE [17], DistMult [20],

---

[1] https://www.google.com/intl/es419/insidesearch/features/search/knowledge.html.

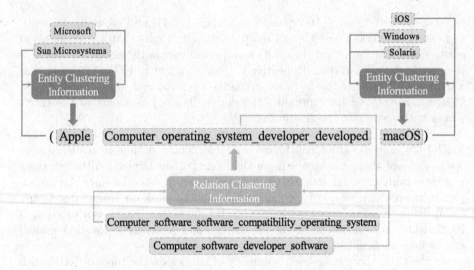

**Fig. 1.** Example of clustering information in freebase

ComplEx [21], RotatE [19], ConvE [23] and DualE [24] to extend. We conduct our experiments on two tasks, including link prediction and triplet classification. The experimental results show that Shared Entity/Relation Cluster embeddings can improve the performance on both two tasks w.r.t different metrics on all datasets. Furthermore, nearly all datasets benefit more from shared entity clustering embeddings due to a large number of entities, leading to more latent related information. Generally, if we use the clustering information of both entities and relations, the models perform better than using only one of them. We highlight our contributions as follows:

1. We propose a clustering-enhanced KGE model. Specifically, we use Shared Entity Cluster (SEC) embeddings, Shared Relation Cluster (SRC) embeddings and Shared Entity & Relation Cluster (SERC) embeddings to utilize the information of similar entities and relations.
2. The proposed method can be applied to many other KGE models. Particularly, we apply the method to six popular KGE models, including TransE, ComplEx, DistMult, RotatE, ConvE and DualE.
3. Extensive experiments on link prediction and triplet classification tasks demonstrate the effectiveness of our method.

## 2   Related Work

In recent years, a lot of works have been conducted on the topic of KGE. Knowledge representation learning roughly falls into two groups. The idea of translational distance has already been used in Word2Vec [25], e.g. *queen - woman ≈ king - man*. TransE [17] is one of the representative models introduced in 2013. After that, there are many variants of TransE, e.g., TransR [26],

TransH [18], TransG [27] and KG2E [28], etc. TransH [18] introduces a relation-specific hyperplane, using a projection matrix to transform entities to target hyperplanes. TransR [26] transforms entities to a relation-specific space with the help of a projection matrix. KG2E [28] uses Gaussian distributions to learn the embeddings. TransG [27] uses a mixture of Gaussian distributions to embed the relations because there are multiple semantics in different relations. Futhermore, there are some models use rotation space to learn the entities and relations, such as RotatE [19], HAKE [29].

Another group is semantic matching models. The main idea of these models is to apply dot multiplication, matrix decomposition, or other operations to evaluate the triplets. Many models, such as RESCAL [30], DistMult [20], ComplEx [21], Holographic Embeddings (HolE) [31] and ANALOGY [32], have been published. The representative model is RESCAL [30], which uses a relation matrix to model the interactions between two entities. DistMult [20] utilizes a diagonal matrix to solve the problem of over-fitting of bilinear model. However, DistMult can not model the asymmetry relations. ComplEx [21] introduces complex-valued embeddings to embed the assymmetry relations. DualE [24] utilizes the dual quaternion number to model different relation patterns. Besides these traditional semantic models, nowadays, there are many models using neural network, such as ConvE [23]. Moreover, some models use graph neural network, such as Relational Graph Convolutional Networks (RGCN) [33] and Relational Graph neural network with Hierarchical Attention (RGHAT) [34], etc. Generative Adversarial Networks are also used in KGC, such as KBGAN [35]. ConvE [23] uses a convolution neural network as a part of the score function. And InteractE [36] makes some improvement based on ConvE. R-GCN [33] uses a relational graph convolution neural network to enhance the expressive power of embeddings. RGHAT [34] introduces a relation convolution neural network with hierarchical attention as an encoder.

What's more, there are some works use extra information. PTransE [37] concentrates on the correlations about continuous paths (relations) on KGs. DKRL [38] combines the latent information of text space and knowledge space, aligns and trains together, while IKRL [39] uses information of image space instead. CTransR [26] tries to find latent correlations for each relation through entities. Zhang et al. [22] introduced a hierarchical relation structure and concentrated on the similarity between different relations. Besides these, there are also some models using the pre-trained language models, i.e., BLP [40] and K-Bert [41]. Most of them try to use latent correlations in relations. However, there are much more entities compared with relations in KGs. We believe there are probably more latent correlations among entities.

In our experiment, we try to find some latent correlations in both relations and entities. We believe that entities also have some similarities because they are extracted from the real world like relations. Similar entities or relations have a relatively short distance in entity or relation space. Moreover, we cluster similar entities and relations to use the related information among them. We propose the shared cluster embeddings in our models to enhance the baseline. The method is able to extend most KGE models. The experimental results show that our models perform better than all the other baselines.

## 3    Problem Description

**Knowledge Graph Embedding (KGE).** Given a knowledge graph $\mathcal{G} = \{(h, r, t)\} \subseteq \mathcal{E} \times \mathcal{R} \times \mathcal{E}$, where $\mathcal{E}$ is the entity set, $\mathcal{R}$ is the relation set, and $(h, r, t)$ is a triplet with head, relation and tail. Knowledge graph embedding aims to embed all the triplets $(h, r, t) \in \mathcal{G}$ to their specific vector $(\mathbf{h}, \mathbf{r}, \mathbf{t})$. And $\mathbf{h}, \mathbf{t}$ are in the entity space $\mathbb{R}^{d_1}$, while $\mathbf{r}$ is in the relation space $\mathbb{R}^{d_2}$, where $d_1$ and $d_2$ are the embedding sizes.

**Clusters in KGs.** There are some entities and relations having the similar semantics in a large scale KG. Similar entities and relations can be grouped into entity clusters $C^{\mathcal{E}} = \{C^{\mathcal{E}_1}, C^{\mathcal{E}_2}, \ldots, C^{\mathcal{E}_{\mathcal{N}_e}}\}$ ($\mathcal{N}_e$ is the number of entity clusters) and relation clusters $C^{\mathcal{R}} = \{C^{\mathcal{R}_1}, C^{\mathcal{R}_2}, \ldots, C^{\mathcal{R}_{\mathcal{N}_r}}\}$ ($\mathcal{N}_r$ is the number of relation clusters).

**Knowledge Graph Embedding with Shared Cluster Information.** In large scale KGs, entity and relation clusters encapsule rich correlations between similar entities and relations. The task of KGE with shared cluster information is to make good use of the cluster information to enhance existing KGE models.

## 4    Methodology

In this section, we will put forward our methods for incorporating cluster embeddings to base models, such as TransE-SEC/SRC/SERC. Then we will present the loss function of our models.

### 4.1    KGE Models with Shared Cluster Embeddings

In hyperplane space, if two entities or relations are close in distance, we suppose they are similar in latent semantic. This will increase the semantic correlations in similar entities and relations. With the aid of shared entity and relation cluster embeddings, the distance between unrelated triplets can be larger than the original one.

Given a dataset, we train a basic model to obtain the pre-trained embeddings of entities and relations. Then we use k-means [42] clustering method to gather the similar embeddings. Suppose pre-trained entity and relation embeddings are $\mathbf{e}_1, \mathbf{e}_2, \ldots, \mathbf{e}_n$ and $\mathbf{r}_1, \mathbf{r}_2, \ldots, \mathbf{r}_m$, where $n$ and $m$ are the number of entities and relations in this dataset. The set of entity clusters created by k-means is $C^{\mathcal{E}} = \{C^{\mathcal{E}_1}, C^{\mathcal{E}_2}, \ldots, C^{\mathcal{E}_{\mathcal{N}_e}}\}$ ($\mathcal{N}_e$ is the number of entity clusters) and relation clusters created by k-means is $C^{\mathcal{R}} = \{C^{\mathcal{R}_1}, C^{\mathcal{R}_2}, \ldots, C^{\mathcal{R}_{\mathcal{N}_r}}\}$ ($\mathcal{N}_r$ is the number of relation clusters). Suppose the set of shared entity cluster embeddings are $\mathbf{c}_1^e, \mathbf{c}_2^e, \ldots, \mathbf{c}_{\mathcal{N}_e}^e$, and shared relation cluster embeddings are $\mathbf{c}_1^r, \mathbf{c}_2^r, \ldots, \mathbf{c}_{\mathcal{N}_r}^r$.

For a given triplet $(h, r, t)$, $\mathbf{h}, \mathbf{r}, \mathbf{t}$ are the low-dimensional embeddings of this triplet. Suppose $\mathbf{h}_c, \mathbf{t}_c$ are shared entity cluster embeddings and $\mathbf{r_c}$ is shared relation cluster embedding. There are three ways to extend:

1. **Shared Entity Cluster embeddings**:

$$\mathbf{h}' = \mathbf{h} + \alpha_e \mathbf{h}_c, \tag{1}$$

$$\mathbf{t}' = \mathbf{t} + \alpha_e \mathbf{t}_c, \tag{2}$$

$\alpha_e$ is the weight of shared entity cluster embeddings. $\mathbf{h}'$ and $\mathbf{t}'$ are new embeddings that combine shared entity cluster embeddings. The larger $\alpha_e$ we use, the more shared information between similar entities the models will learn while training.

2. **Shared Relation Cluster embeddings**:

$$\mathbf{r}' = \mathbf{r} + \alpha_r \mathbf{r}_c, \tag{3}$$

$\alpha_r$ is the weight of shared relation cluster embeddings. $\mathbf{r}'$ is a new embedding that combines shared relation cluster embeddings.The larger $\alpha_r$ we use, the more shared information between similar relations the models will learn while training.

3. **Shared Entity & Relation Cluster embeddings**:
   This model combines both shared entity and relation cluster embeddings above, which is shown as Eqs. (1–3).

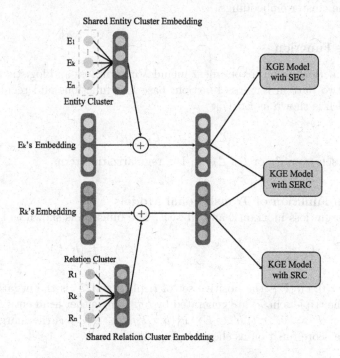

**Fig. 2.** Architecture of our models.

Figure 2 shows the architecture of our models. We use TransE as an example. TransE-SEC is the model with entity cluster embeddings. TransE-SRC is the model with relation cluster embeddings, and TransE-SERC is the model with both of them.

The score functions of TransE-SEC, TransE-SRC and TransE-SERC are shown as Eqs. (4–6):

1. TransE-SEC:

$$f_r(h,t) = \|(\mathbf{h} + \alpha_e \mathbf{h}_c) + \mathbf{r} - (\mathbf{t} + \alpha_e \mathbf{t}_c)\|_{L_n}, \tag{4}$$

2. TransE-SRC:

$$f_r(h,t) = \|\mathbf{h} + (\mathbf{r} + \alpha_r \mathbf{r}_c) - \mathbf{t}\|_{L_n}, \tag{5}$$

3. TransE-SERC:

$$f_r(h,t) = \|(\mathbf{h} + \alpha_e \mathbf{h}_c) + (\mathbf{r} + \alpha_r \mathbf{r}_c) - (\mathbf{t} + \alpha_e \mathbf{t}_c)\|_{L_n}, \tag{6}$$

where $\mathbf{h}_c, \mathbf{t}_c$ are shared entity cluster embeddings of $h$ and $t$, and $\mathbf{r}_c$ is shared relation cluster embedding. $L_n$ can be $L_1$ or $L_2$ norm.

As for other models, both of them have three types of extended models. ComplEx also has imaginary cluster embeddings and DualE also has other imaginary parts of the cluster embeddings.

## 4.2    Loss Function

The loss functions in our experiment includes margin loss and logistic loss. And there are two part in our loss function: base loss functions and regularization terms, which is shown as Eq. (7):

$$\mathcal{L} = \mathcal{L}_{a/b} + \mathcal{L}_{Regul}, \tag{7}$$

$\mathcal{L}_{a/b}$ is chosen from $\mathcal{L}_a$ or $\mathcal{L}_b$, $\mathcal{L}_{Regul}$ is a regularization term.

### 4.2.1 Loss Function of Translational Models

We use margin loss in TransE and RotatE. The function is shown as Eq. (8):

$$\mathcal{L}_a = \sum_{(h,r,t)\in \mathcal{T}, (h',r,t')\in \mathcal{T}'} [\gamma + f_r(h,t) - f_r(h',t')]_+, \tag{8}$$

$[x]_+ = max(0, x)$, $\mathcal{T}$ is the positive set of triplets and $\mathcal{T}'$ is the negative set of triplets. The triplets in $\mathcal{T}'$ are generated by replacing either head entity $h$ or tail entity $t$, i.e. $\mathcal{T}' = \{h', r, t) | h' \in E\} \cup \{(h, r, t') | t' \in E\}$. $\gamma$ is the margin value. $f_r(h,t)$ is a score function as shown in Eqs. (4–6).

### 4.2.2 Loss Function of Other Models

For the other models (Distmult, ComplEx, ConvE and DualE), we use logistic loss function, which is shown as Eq. (9):

$$\mathcal{L}_b = \sum_{(h,r,t)\in\mathcal{T}\cup\mathcal{T}'} \log\left(1 + \exp\left(-y_{hrt}\cdot f_r(h,t)\right)\right). \tag{9}$$

If $(h,r,t) \in \mathcal{T}$, $y_{hrt} = 1$, otherwise $y_{hrt} = -1$. $\mathcal{T}$ is the positive set of triplets and $\mathcal{T}'$ is the negative set of triplets.

### 4.2.3 Regularization Terms

There are three regularization terms in both margin loss and logistic loss, as is shown in Eq. (10):

$$\mathcal{L}_{Regul} = \sum_{(h,r,t)\in\mathcal{T}} \left[\lambda_1(\|\mathbf{h}\|_2^2 + \|\mathbf{r}\|_2^2 + \|\mathbf{t}\|_2^2) + \lambda_2(\|\mathbf{h}_c\|_2^2 + \|\mathbf{t}_c\|_2^2) + \lambda_3\|\mathbf{r}_c\|_2^2\right],$$
$$\tag{10}$$

where $\mathbf{h}_c$ and $\mathbf{t}_c$ are the corresponding entity cluster embeddings of $h$ and $t$. And $\mathbf{r}_c$ is relation cluster embedding of $r$.

$\lambda_1, \lambda_2$ and $\lambda_3$ are trade-off parameters to balance the basic embeddings and cluster embeddings. Large value of $\lambda_1$ results in entities and relations in the same clusters sharing the same embedding vector. And large value of $\lambda_2$ and $\lambda_3$ leads to learning separate shared cluster embedding vector while training.

## 5    Experiment

We introduce the data sets, baseline models, experimental results and analysis of hyper-parameters in this section.

### 5.1    Data Sets

We use two widely used benchmarks on link prediction to evaluate our extended models, including FB15k-237 and WN18RR. For triplet classification, we use FB13 and WN11 datasets. FB15k-237 and FB13 are popular datasets which are extracted from Freebase. Unlike FB15k and WN18, they do not include reverse relations in the dataset. WN18RR and WN11 are extracted from WordNet. The statistics are shown as Table 1:

Table 1. Statistics of experimental datasets.

| Dataset | $|\mathcal{E}|$ | $|\mathcal{R}|$ | Train | Valid | Test |
|---------|------|------|---------|--------|--------|
| FB15k-237 | 14,541 | 237 | 272,115 | 17,535 | 20,466 |
| WN18RR | 40,943 | 11 | 86,835 | 3,034 | 3,134 |
| FB13 | 75,043 | 13 | 316,232 | 5,908 | 23,733 |
| WN11 | 38,696 | 11 | 112,581 | 2,609 | 10,544 |

## 5.2  Baseline

To demonstrate the effectiveness of our models, we choose some baselines which are widely used to compare:

1. **TransE** [17]: the model is a translational distance modelh.
2. **DistMult** [20]: a semantic matching model, which uses an efficient bilinear function as the score function.
3. **ComplEx** [21]: the model extends the embeddings from the real space to the complex space, which utilizes richer information.
4. **RotatE** [19]: a translational distance model, which uses a rotation operation of relation in complex vector space.
5. **ConvE** [23]: the model uses a convolution neural network as a part of the score function.
6. **DualE** [24]: the model utilizes the dual quaternion number to model different relation patterns.

## 5.3  Link Prediction

Link Prediction is to complete the missing entity $h$ or $t$ while giving an incomplete triplet $(h, r, ?)$ or $(?, r, t)$.

### 5.3.1 Parameters and Metrics

We implement TransE, DistMult, ComplEx, RotatE, ConvE and DualE for comparison and use Adam [43] optimizer in our experiments. To compare fairly for all models, the embedding dimension $dim$ of all models on FB15k-237 is 100, and the embedding dimension on WN18RR is 200. Training epochs for all models are set to 600. For the entity and relation clusters, we first run every model to get the embeddings of entities and relations. Then we obtain the clusters with k-means algorithm, respectively. The detailed parameters are shown in Table 7.

In the test steps, we replace the head entity or tail entity of a test case with all entities in KG and rank all the scores obtained from our models. And then, we calculate the mean value of metrics as final results. After that, we can get three values: (1) Mean Reciprocal Rank (MRR), the mean reciprocal of all the predicted ranks; (2) Mean Rank (MR), the mean of all the predicted ranks; (3) H@n, the proportion of ranks not larger than n. As we know, lower MR, larger MRR and larger H@n indicate better performance.

**Table 2.** Experimental results on FB15k-237 and WN18RR.

| Model | FB15k-237 | | | | | WN18RR | | | | |
|---|---|---|---|---|---|---|---|---|---|---|
| | MR | MRR | H@10 | H@3 | H@1 | MR | MRR | H@10 | H@3 | H@1 |
| R-GCN+ [24] | - | 0.249 | 41.7 | 26.4 | 15.1 | – | – | – | – | – |
| ConvKB [44] | 216 | 0.289 | 47.1 | 32.4 | 19.8 | <u>1295</u> | 0.265 | 55.8 | 44.5 | 5.8 |
| TransE | 177 | 0.289 | 46.24 | 31.59 | 20.23 | 2916 | 0.209 | 47.94 | 33.36 | 5.08 |
| TransE-SEC | 168 | 0.298 | 47.64 | 32.76 | 20.75 | 2857 | 0.212 | 48.58 | 33.67 | **5.14** |
| TransE-SRC | 173 | 0.293 | 46.54 | 32.16 | 20.61 | 2960 | 0.211 | 48.47 | 33.60 | 5.03 |
| TransE-SERC | **166** | **0.301** | **47.81** | **33.08** | **21.19** | **2855** | **0.213** | **48.78** | **34.23** | 5.03 |
| DistMult | 248 | 0.277 | 44.31 | 30.32 | 19.37 | 5405 | 0.429 | 47.76 | 43.59 | 40.43 |
| DistMult-SEC | **213** | 0.291 | 46.04 | **32.03** | 20.53 | 5255 | 0.445 | 49.80 | 45.58 | **41.74** |
| DistMult-SRC | 241 | 0.279 | 44.46 | 30.76 | 19.59 | **4905** | 0.436 | 48.85 | 44.87 | 40.80 |
| DistMult-SERC | 231 | **0.292** | **46.28** | 31.90 | **20.61** | 5173 | **0.446** | **50.21** | **45.73** | 41.71 |
| ComplEx | 232 | 0.275 | 45.03 | 30.57 | 18.73 | 6387 | 0.441 | 48.32 | 44.89 | 41.99 |
| ComplEx-SEC | 218 | 0.291 | 46.42 | 32.05 | 20.40 | 5377 | 0.464 | **51.58** | **47.61** | 43.70 |
| ComplEx-SRC | 223 | 0.283 | 45.18 | 31.13 | 19.89 | 6013 | 0.446 | 49.09 | 45.20 | 42.37 |
| ComplEx-SERC | **188** | **0.293** | **47.00** | **32.24** | **20.60** | **5102** | **0.466** | 51.42 | 47.54 | **43.95** |
| RotatE | 285 | 0.293 | 47.12 | 32.45 | 20.53 | 4607 | 0.471 | 55.13 | 48.83 | 42.99 |
| RotatE-SEC | 203 | 0.309 | 49.79 | 34.51 | 21.47 | 4107 | 0.477 | 55.95 | 49.38 | 43.46 |
| RotatE-SRC | 230 | 0.303 | 48.31 | 33.41 | 21.35 | 4741 | 0.475 | 55.16 | 48.78 | 43.80 |
| RotatE-SERC | **198** | **0.312** | **49.82** | **34.78** | **22.00** | **3960** | **0.479** | **56.05** | **49.59** | **43.99** |
| ConvE | 207 | 0.276 | 44.52 | 30.36 | 19.22 | 4308 | 0.427 | 49.06 | 43.98 | 39.37 |
| ConvE-SEC | 191 | 0.289 | 46.54 | 31.65 | 20.37 | 3814 | 0.436 | 50.59 | 45.01 | 40.11 |
| ConvE-SRC | 202 | 0.281 | 45.93 | 30.78 | 19.89 | 4129 | 0.430 | 49.68 | 44.52 | 39.56 |
| ConvE-SERC | **183** | **0.293** | **46.89** | **32.00** | **20.79** | **3681** | **0.439** | **51.02** | **45.38** | **40.40** |
| DualE | 185 | 0.330 | 51.67 | 36.28 | 23.86 | 3585 | 0.479 | 56.66 | 50.17 | 43.07 |
| DualE-SEC | 176 | 0.341 | 53.54 | 37.31 | 24.99 | 3262 | 0.488 | 57.93 | 51.15 | 43.97 |
| DualE-SRC | 181 | 0.334 | 52.26 | 36.61 | 24.14 | 3493 | 0.483 | 57.04 | 50.65 | 43.52 |
| DualE-SERC | **169** | <u>**0.344**</u> | <u>**54.06**</u> | <u>**37.56**</u> | <u>**25.22**</u> | 3157 | <u>**0.491**</u> | <u>**58.30**</u> | <u>**51.46**</u> | <u>**44.36**</u> |

**Table 3.** Examples of entity clusters in FB15k-237

| | Entities |
|---|---|
| 1 | Academy Award for Best Foreign Language Film, Golden Globe Award for Best Foreign Language Film |
| 2 | Italian Language, English Language, French Language, Russian Language $\cdots$ |
| 3 | Windows XP, Windows Vista |
| 4 | Jiangsu, Suzhou, Nanjing |

## 5.3.2 Experiment Results

In this section, we introduce the evaluation results w.r.t link prediction metrics and case studies.

**Table 4.** Examples of relation clusters in FB15k-237

| | Relations |
|---|---|
| 1 | /film/film/language, /tv/tv_program/languages |
| 2 | /people/person/profession, /people/profession/specialization_of |
| 3 | /location/statistical_region/religions./location/religion_percentage/religion, /people/person/religion |

**Table 5.** Examples of entity clusters in WN18RR

| | Entities |
|---|---|
| 1 | queen_regnant, prince, queen, princess, highness |
| 2 | Spread, circulation, pass_on |
| 3 | Vary, variant, variation, deviation |
| 4 | Traduce, vilification, vilifier |

### *Evaluation Results*

Table 2 shows the link prediction results. We divide the table into seven groups. The first group is the other baselines that we do not extend. Each of the latter six groups contains a baseline and three extended models, including models with Shared Entity Cluster (SEC) embeddings, models with Shared Relation Cluster(SRC) embeddings and models with Shared Entity & Relation Cluster(SERC) embeddings. The results in bold font are the best ones in every groups, and the underlined results are the best ones in the whole column. From the results, we get these conclusions below:

1. Generally, our extended models outperform the baselines. This indicates that the shared cluster embeddings is capable of providing fruitful information and have a positive effect to enhance KGE models.
2. The models with SERC obtain a better result than the ones with SEC, and models with SEC obtain a better result than the ones with SRC, which implies that the combination of the entity cluster information and relation cluster information is helpful for representation learning. Because there are much more entities compared with relations, the entities that join the same cluster have richer semantic correlations than relations. Thus the information learned from SEC is more valuable than that learned from SRC.
3. Based on the best basic model DualE, DualE-SERC achieves the best results on both datasets.

## Case Study

We also provide some case studies on entity clusters and relation clusters. Table 3 shows some entity clusters of FB15k-237. Cluster 1 to 4 includes some entities about the film awards, languages, operating systems, and cities in China. Table 4 shows some relation clusters of FB15k-237. The contents of these clusters are some media languages, the profession of people, and religion. The related relations also have been gathered together. Both entity clusters and relation clusters enrich the models' representation. Table 5 shows some entity clusters of WN18RR. Moreover, again, Cluster 1 to 4 gathers entities about members of the Royal family, synonyms about the word "spread", some variants about the word "vary" and some synonyms about the word "traduce". We can see that similar, semantic-related entities join the same cluster.

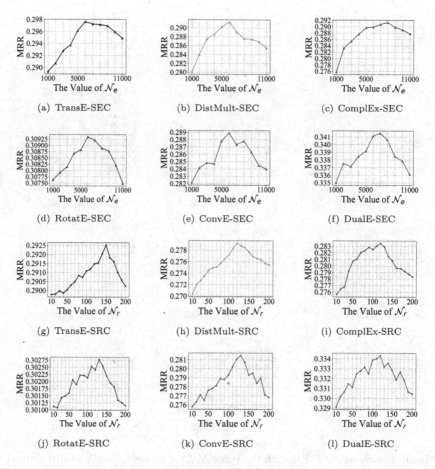

**Fig. 3.** The change of MRR with the value $\mathcal{N}_e/\mathcal{N}_r$ increasing on FB15k-237 dataset

### 5.3.3 Parameter Study

In this section, we study the influence that is affected by the number of entity/relation clusters $\mathcal{N}_e/\mathcal{N}_r$ and the weight of shared entity/relation embeddings $\alpha_e/\alpha_r$.

**Parameter 1: Entity/Relation cluster number $\mathcal{N}_e/\mathcal{N}_r$**

We set a fixed interval for the experimental process. For the cluster number of entities $\mathcal{N}_e$ on FB15k-237, we vary it from 1000 to 11000, and the interval is 1000. For the cluster number of relations $\mathcal{N}_r$, we set it from 10 to 200, and the interval is 10. As for WN18RR, we set entity cluster number $\mathcal{N}_e$ from 1000 to 33000, and the interval is 4000 while relation cluster number $\mathcal{N}_r$ is from 2 to 9 one by one.

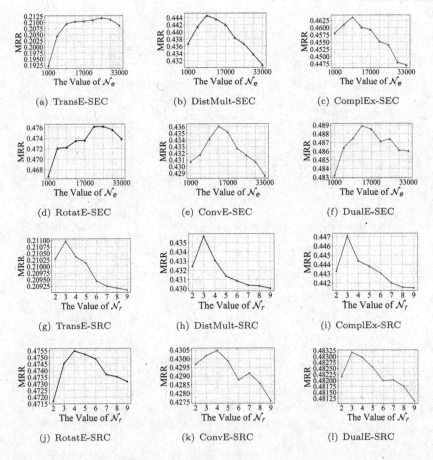

**Fig. 4.** The change of MRR with the value $\mathcal{N}_e/\mathcal{N}_r$ increasing on WN18RR dataset

Figure 3 shows all the experimental results on FB15k-237 dataset w.r.t MRR metrics with different clustering numbers. And the results on WN18RR dataset are shown in Fig. 4. All the six models have an optimal $\mathcal{N}_e$ and $\mathcal{N}_r$ on both FB15k-237 and WN18RR datasets. The MRR goes up when the $\mathcal{N}_e/\mathcal{N}_r$ increases to the optimal value. After the optimal value of $\mathcal{N}_e$ and $\mathcal{N}_r$, the results of MRR fall down. This reason lies as the small value of $\mathcal{N}_e/\mathcal{N}_r$ leads to large-sized clusters. Some unrelated entities/relations may join the same cluster, thus they degrade the performance. A large value of $\mathcal{N}_e/\mathcal{N}_r$ leads to small-sized clusters. In this case, each cluster only has a few entities/relations, thus the information leveraged by each entity/relation is limited, leading to unsatisfactory results.

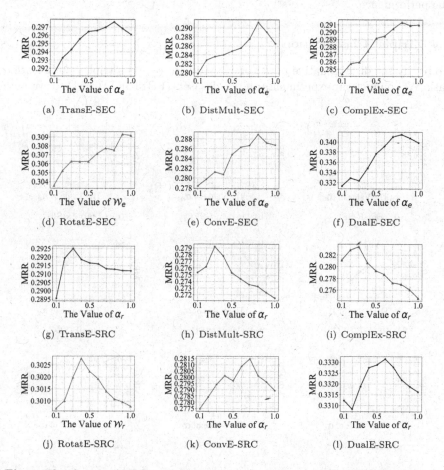

**Fig. 5.** The change of MRR with the value $\alpha_e/\alpha_r$ increasing on FB15k-237 dataset

## Parameter 2: Weight of Shared Entity/Relation Cluster embedding $\alpha_e/\alpha_r$

In this experiment, we limit the value $\alpha_e/\alpha_r$ from 0.1 to 1.0. And the interval is set to 0.1.

Figures 5 and 6 represent the performance of six models on FB15k-237 and WN18RR datasets. From these Figures, we can see that both SEC and SRC have an optimal value of $\alpha_e$ and $\alpha_r$. Different models, i.e., TransE and DistMult, have an approximate optimal value. The MRR comes to the peak with the increasing weight of $\alpha_e/\alpha_r$. And after the $\alpha_e/\alpha_r$ becomes larger than the optimal one, the metric of MRR falls down. We can conclude that small weight of $\alpha_e/\alpha_r$ leads to limited shared information in the process of learning the embeddings, which will make the performance worse than the optimal result (better than baseline). Larger weight of $\alpha_e/\alpha_r$ leads to too much shared information, which overrides the importance of original entity/relation embeddings and degrades the performance.

## 5.4   Triplet Classification

To testify the discriminative power of our models, we add a triplet classification task. We conduct our experiments on two datasets (FB13 and WN11) to evaluate

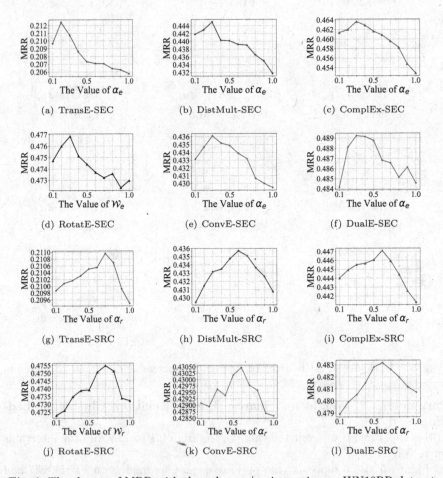

**Fig. 6.** The change of MRR with the value $\alpha_e/\alpha_r$ increasing on WN18RR dataset

**Table 6.** Triplet classification results on FB13 and WN11.

| Model | FB13 | WN11 | AVG |
|---|---|---|---|
| TransE | 81.5 | 75.9 | 78.7 |
| TransE-SEC | 85.3 | **86.5** | 85.9 |
| TransE-SRC | 82.1 | 85.6 | 83.9 |
| TransE-SERC | **86.1** | **86.5** | **86.3** |
| DistMult | 84.9 | 76.8 | 80.9 |
| DistMult-SEC | 85.6 | 84.3 | 85.0 |
| DistMult-SRC | 85.7 | 79.8 | 82.8 |
| DistMult-SERC | **86.0** | **84.8** | **85.4** |
| ComplEx | 85.6 | 76.5 | 81.1 |
| ComplEx-SEC | 86.1 | 84.4 | 85.3 |
| ComplEx-SRC | 86.0 | 79.3 | 82.7 |
| ComplEx-SERC | **86.5** | **84.9** | **85.7** |
| RotatE | 86.1 | 82.1 | 84.2 |
| RotatE-SEC | 87.4 | 85.3 | 86.4 |
| RotatE-SRC | 86.5 | 82.9 | 84.7 |
| RotatE-SERC | **87.9** | **86.0** | **87.0** |
| ConvE | 85.9 | 84.7 | 85.3 |
| ConvE-SEC | 87.3 | 86.0 | 86.7 |
| ConvE-SRC | 86.2 | 85.5 | 85.9 |
| ConvE-SERC | **87.5** | **86.1** | **86.8** |
| DualE | 86.9 | 84.0 | 85.5 |
| DualE-SEC | 88.0 | 85.9 | 87.0 |
| DualE-SRC | 87.2 | 84.6 | 85.9 |
| DualE-SERC | **88.6** | **86.5** | **87.6** |

our models. And we adopt accuracy of classification as the metric for comparison. We follow the decision process, which is first introduced by NTN [45]: for each relation, we find a threshold $T_R$. If $f_r(h, r, t) \geq T_R$, the triplet $(h, r, t)$ holds, otherwise it does not hold.

The experimental results are shown in Table 6. Bold fonts represent the best result in the group. Table 6 are divided into six groups. Every group includes a baseline and their extended models. From Table 6, we observe that all of our models outperform the baselines. For all groups, models with both entity and relation cluster embeddings, such as TransE-SERC, perform the best.

## 6 Conclusion

In this paper, we proposed to use shared clustering embeddings to enhance KGE models. In this way, each entity or relation is able to leverage the valuable information from semantically similar entities or relations. Our method is capable of extending various KGE models. Particularly, we extend popular KGE model, e.g., TransE, DistMult, ComplEx, RotatE, ConvE and DualE. Finally, extensive experiments on popular benchmarks validated the powerful ability of the proposed method.

**Acknowledgements.** The research work supported by the National Natural Science Foundation of China under Grant No.61976204, U1811461, U1836206.

## Appendix A    Model parameters

**Table 7.** Model parameters. E,R and E&R are the abbreviations of Entity clustering, Relation clustering and Entity&Relation clustering, respectively.

| Base model | Dataset | Type | $lr$ | $\gamma$ | $dim$ | $\mathcal{N}_e$ | $\mathcal{N}_r$ | $\alpha_e$ | $\alpha_r$ | $\lambda_1$ | $\lambda_2$ | $\lambda_3$ |
|---|---|---|---|---|---|---|---|---|---|---|---|---|
| TransE | FB15k-237 | E | 0.003 | 1.0 | 100 | 6000 | - | 0.8 | - | 1e-8 | 1e-9 | - |
| | | R | 0.003 | 1.0 | 100 | - | 150 | - | 0.3 | 1e-8 | - | 1e-9 |
| | | E&R | 0.003 | 1.0 | 100 | 6000 | 150 | 150 | 0.3 | 1e-8 | 1e-9 | 1e-9 |
| | WN18RR | E | 0.03 | 1.0 | 200 | 28000 | - | 0.2 | - | 1e-10 | 1e-10 | - |
| | | R | 0.03 | 1.0 | 200 | - | 3 | - | 0.7 | 1e-10 | - | 1e-10 |
| | | E&R | 0.03 | 1.0 | 200 | 28000 | 3 | 0.2 | 0.7 | 1e-10 | 1e-10 | 1e-10 |
| DistMult | FB15k-237 | E | 0.01 | - | 100 | 6000 | - | 0.8 | - | 1e-9 | 1e-9 | - |
| | | R | 0.01 | - | 100 | - | 120 | - | 0.3 | 1e-9 | - | 1e-9 |
| | | E&R | 0.01 | - | 100 | 6000 | 120 | 0.8 | 0.3 | 1e-9 | 1e-9 | 1e-9 |
| | WN18RR | E | 0.01 | - | 200 | 9000 | - | 0.3 | - | 1e-10 | 1e-10 | - |
| | | R | 0.01 | - | 200 | - | 3 | - | 0.6 | 1e-10 | - | 1e-10 |
| | | E&R | 0.01 | - | 200 | 9000 | 3 | 0.3 | 0.6 | 1e-10 | 1e-10 | 1e-10 |
| ComplEx | FB15k-237 | E | 0.01 | - | 100 | 8000 | - | 0.8 | - | 1e-9 | 1e-9 | - |
| | | R | 0.01 | - | 100 | - | 120 | - | 0.3 | 1e-9 | - | 1e-9 |
| | | E&R | 0.01 | - | 100 | 8000 | 120 | 0.8 | 0.3 | 1e-9 | 1e-9 | 1e-9 |
| | WN18RR | E | 0.01 | - | 200 | 9000 | - | 0.3 | - | 1e-10 | 1e-10 | - |
| | | R | 0.01 | - | 200 | - | 3 | - | 0.6 | 1e-10 | - | 1e-10 |
| | | E&R | 0.01 | - | 200 | 9000 | 3 | 0.3 | 0.6 | 1e-10 | 1e-10 | 1e-10 |
| RotatE | FB15k-237 | E | 5e-4 | 12.0 | 100 | 6000 | - | 0.9 | - | 1e-7 | 1e-8 | - |
| | | R | 5e-4 | 12.0 | 100 | - | 130 | - | 0.4 | 1e-7 | - | 1e-8 |
| | | E&R | 5e-4 | 12.0 | 100 | 6000 | 130 | 0.9 | 0.4 | 1e-7 | 1e-8 | 1e-8 |
| | WN18RR | E | 5e-4 | 3.0 | 200 | 27000 | - | 0.3 | - | 1e-9 | 1e-10 | - |
| | | R | 5e-4 | 3.0 | 200 | - | 4 | - | 0.7 | 1e-9 | - | 1e-10 |
| | | E&R | 5e-4 | 3.0 | 200 | 27000 | 4 | 0.3 | 0.7 | 1e-9 | 1e-10 | 1e-10 |
| DualE | FB15k-237 | E | 0.01 | - | 100 | 7000 | - | 0.8 | - | 1e-9 | 1e-9 | - |
| | | R | 0.01 | - | 100 | - | 120 | - | 0.6 | 1e-9 | - | 1e-9 |
| | | E&R | 0.01 | - | 100 | 7000 | 120 | 0.8 | 0.6 | 1e-9 | 1e-9 | 1e-9 |
| | WN18RR | E | 0.01 | - | 200 | 9000 | - | 0.3 | - | 1e-10 | 1e-10 | - |
| | | R | 0.01 | - | 200 | - | 3 | - | 0.6 | 1e-10 | - | 1e-10 |
| | | E&R | 0.01 | - | 200 | 9000 | 3 | 0.3 | 0.6 | 1e-10 | 1e-10 | 1e-10 |
| ConvE | FB15k-237 | E | 0.04 | - | 100 | 6000 | - | 0.8 | - | 1e-9 | 1e-9 | - |
| | | R | 0.04 | - | 100 | - | 130 | - | 0.7 | 1e-9 | - | 1e-9 |
| | | E&R | 0.04 | - | 100 | 6000 | 130 | 0.8 | 0.7 | 1e-9 | 1e-9 | 1e-9 |
| | WN18RR | E | 0.01 | - | 200 | 9000 | - | 0.3 | - | 1e-10 | 1e-10 | - |
| | | R | 0.01 | - | 200 | - | 4 | - | 0.6 | 1e-10 | - | 1e-10 |
| | | E&R | 0.01 | - | 200 | 9000 | 4 | 0.3 | 0.6 | 1e-10 | 1e-10 | 1e-10 |

# References

1. Hoffmann, R., Zhang, C., Ling, X., Zettlemoyer, L., Weld, D.S.: Knowledge-based weak supervision for information extraction of overlapping relations. In: Annual Meeting of the Association for Computational Linguistics (2011)
2. Wang, Q., Mao, Z., Wang, B., Guo, L.: Knowledge graph embedding: a survey of approaches and applications. IEEE Trans. Knowl. Data Eng. **29**, 2724–2743 (2017)
3. Li, Z., Sun, Y., Tang, S., Zhang, C., Ma, H.: Adaptive graph convolutional networks with attention mechanism for relation extraction. In: International Joint Conference on Neural Networks (2020)
4. Li, Z., Sun, Y., Zhu, J., Tang, S., Zhang, C., Ma, H.: Improve relation extraction with dual attention-guided graph convolutional networks. Neural Computing and Applications (2021)
5. Riedel, S., Yao, L., McCallum, A., Marlin, B.M.: Relation extraction with matrix factorization and universal schemas. In: Conference of the North American Chapter of the Association for Computational Linguistics (2013)
6. Weston, J., Bordes, A., Yakhnenko, O., Usunier, N.: Connecting language and knowledge bases with embedding models for relation extraction. In: Conference on Empirical Methods in Natural Language Processing (2013)
7. Koren, Y., Bell, R., Volinsky, C.: Matrix factorization techniques for recommender systems. Computer. **42**, 30–37 (2009)
8. Wang, X., Huang, T., Wang, D., Yuan, Y., Liu, Z., He, X., Chua, T.-S.: Learning intents behind interactions with knowledge graph for recommendation. In: Proceedings of the Web Conference (2021)
9. Yu, X., et al.: Personalized entity recommendation: a heterogeneous information network approach. In: ACM International Conference on Web Search and Data Mining, pp. 283–292 (2014)
10. Saxena, A., Tripathi, A., Talukdar, P.: Improving multi-hop question answering over knowledge graphs using knowledge base embeddings. In: Annual Meeting of the Association for Computational Linguistics (2020)
11. Bordes, A., Weston, J., Usunier, N.: Open question answering with weakly supervised embedding models. In: Calders, T., Esposito, F., Hüllermeier, E., Meo, R. (eds.) ECML PKDD 2014. LNCS (LNAI), vol. 8724, pp. 165–180. Springer, Heidelberg (2014). https://doi.org/10.1007/978-3-662-44848-9_11
12. Wang, D., et al.: Cross-lingual knowledge transferring by structural correspondence and space transfer. IEEE Trans. Cybern. **52**, 1–11 (2021)
13. Bollacker, K., Evans, C., Paritosh, P., Sturge, T., Taylor, J.: Freebase: a collaboratively created graph database for structuring human knowledge. In: ACM Conference on Management of Data (2008)
14. Miller, G.A.: Wordnet: a lexical database for English. Commun. ACM. **38**, 39–41 (1995)
15. Lehmann, J., et al.: DBPedia-a large-scale, multilingual knowledge base extracted from Wikipedia. Semant. Web. **6**, 169–175 (2015)
16. Suchanek, F.M., Kasneci, G., Weikum, G.: YAGO: a core of semantic knowledge. In: International World Wide Web Conferences (2007)
17. Bordes, A., Usunier, N., Garcia-Duran, A., Weston, J., Yakhnenko, O.: Translating embeddings for modeling multi-relational data. In: Annual Conference on Neural Information Processing Systems (2013)
18. Wang, Z., Zhang, J., Feng, J., Chen, Z.: Knowledge graph embedding by translating on hyperplanes. In: AAAI Conference on Artificial Intelligence (2014)

19. Sun, Z., Deng, Z.-H., Nie, J.-Y., Tang, J.: Rotate: Knowledge graph embedding by relational rotation in complex space. In International Conference on Learning Representations (2019)
20. Yang, B., Yih, W.-T., He, X., Gao, J., Deng, L.: Embedding entities and relations for learning and inference in knowledge bases. In: International Conference on Learning Representations (2014)
21. Trouillon, T., Welbl, J., Riedel, S., Gaussier, É., Bouchard, G.: Complex embeddings for simple link prediction. In: International Conference on Machine Learning (2016)
22. Zhang, Z., Zhuang, F., Qu, M., Lin, F., He, Q.: Knowledge graph embedding with hierarchical relation structure. In: Conference on Empirical Methods in Natural Language Processing (2018)
23. Dettmers, T., Minervini, P., Stenetorp, P., Riedel, S.: Convolutional 2d knowledge graph embeddings. In: AAAI Conference on Artificial Intelligence (2018)
24. Cao, Z., Xu, Q., Yang, Z., Cao, X., Huang, Q.: Dual quaternion knowledge graph embeddings. In: AAAI Conference on Artificial Intelligence (2021)
25. Mikolov, T., Chen, K., Corrado, G., Dean, J.: Efficient estimation of word representations in vector space. In: International Conference on Learning Representations (2013)
26. Lin, Y., Liu, Z., Sun, M., Liu, Y., Zhu, X.: Learning entity and relation embeddings for knowledge graph completion. In: AAAI Conference on Artificial Intelligence (2015)
27. Xiao, H., Huang, M., Zhu, X.: TranSG: a generative model for knowledge graph embedding. In: Annual Meeting of the Association for Computational Linguistics (2016)
28. He, S., Liu, K., Ji, G., Zhao, J.: Learning to represent knowlergcndge graphs with gaussian embedding. In: ACM International Conference on Information and Knowledge Management (2015)
29. Zhang, Z., Cai, J., Zhang, Y., Wang, J.: Learning hierarchy-aware knowledge graph embeddings for link prediction. In: AAAI Conference on Artificial Intelligence (2020)
30. Nickel, M., Tresp, V., Kriegel, H.-P.: A three-way model for collective learning on multi-relational data. In: International Conference on Machine Learning (2011)
31. Nickel, M., Rosasco, L., Poggio, T.A., et al: Holographic embeddings of knowledge graphs. In: AAAI Conference on Artificial Intelligence (2016)
32. Liu, H., Wu, Y., Yang, Y.: Analogical inference for multi-relational embeddings. In: International Conference on Machine Learning (2017)
33. Schlichtkrull, M., Kipf, T.N., Bloem, P., van den Berg, R., Titov, I., Welling, M.: Modeling relational data with graph convolutional networks. In: Gangemi, A., et al. (eds.) ESWC 2018. LNCS, vol. 10843, pp. 593–607. Springer, Cham (2018). https://doi.org/10.1007/978-3-319-93417-4_38
34. Zhang, Z., Zhuang, F., Zhu, H., Shi, Z.-P., Xiong, H., He, Q.: Relational graph neural network with hierarchical attention for knowledge graph completion. In: AAAI Conference on Artificial Intelligence (2020)
35. Cai, L., Wang, W.Y.: KBGAN: adversarial learning for knowledge graph embeddings. In: Conference of the North American Chapter of the Association for Computational Linguistics (2017)
36. Vashishth, S., Sanyal, S., Nitin, V., Agrawal, N., Talukdar, P.: Interact: improving convolution-based knowledge graph embeddings by increasing feature interactions. In: AAAI Conference on Artificial Intelligence (2020)

37. Lin, Y., Liu, Z., Luan, H., Sun, M., Rao, S., Liu, S.: Modeling relation paths for representation learning of knowledge bases. In: Conference on Empirical Methods in Natural Language Processing (2015)
38. Xie, R., Liu, Z., Jia, J., Luan, H., Sun, M.: Representation learning of knowledge graphs with entity descriptions. In: AAAI Conference on Artificial Intelligence (2016)
39. Xie, R., Liu, Z., Luan, H., Sun, M.: Image-embodied knowledge representation learning. In: International Joint Conference on Artificial Intelligence (2016)
40. Daza, D., Cochez, M., Groth, P.: Inductive entity representations from text via link prediction. In: the Web Conference (2021)
41. Liu, W., et al.: K-BERT: enabling language representation with knowledge graph. In: AAAI Conference on Artificial Intelligence (2020)
42. MacQueen, J., et al.: Some methods for classification and analysis of multivariate observations. In: Berkeley Symposium on Mathematical Statistics and Probability (1967)
43. Kingma, D.P., Ba, J.: Adam: a method for stochastic optimization. In: International Conference on Learning Representations (2014)
44. Dai Quoc Nguyen, T.D.N., Nguyen, D.Q., Phung, D.: A novel embedding model for knowledge base completion based on convolutional neural network. In: Proceedings of NAACL-HLT, pp. 327–333 (2018)
45. Socher, R., Chen, D., Manning, C.D., Ng, A.: Reasoning with neural tensor networks for knowledge base completion. In: Annual Conference on Neural Information Processing Systems (2013)

# FCI: Feature Cross and User Interest Network

Dejun Lei[✉]

School of Computer Science, Southwest Petroleum University, No. 8, Xindu Avenue,
Xindu District, Chengdu 610599, Sichuan, China
leidejun1992@sina.com

**Abstract.** Recommendation systems are widely used on Internet platforms and the Recommendation system is one of the most mature scenarios for data mining applications. Ranking algorithms in recommender
systems are the most effective means to increase user clicks. At present,
the recommendation system ranking algorithm is mainly divided into
two categories. One is based on data mining, such as user features, product features, and scene features and crosses these functions manually
or automatically. The second method, by mining the user's behavior
sequence, obtains the user's interests and displays the product to the
user. The first approach does not mine useful behavioral data. The second method cannot effectively perform feature crossover and loses some
crossover information between users and products. This paper proposes
a new ranking method for recommender systems. Feature Cross and User
Interest Network FCI. It can efficiently and automatically discover cross
features based on users, products, and scenarios. At the same time, mining the historical behavior information of users. In this algorithm, firstly
our put-forward model resembles a wide-depth structure. Wide structure puts feature intersection network, deep structure puts user historical behavior mining network. Secondly experimental comparisons are
made by changing different feature intersection networks and user behavior sequence networks. Finally, by adding the pre-training vector of the
product, compare different pre-training methods. This approach greatly
improves the CTR in practical recommender systems.

**Keywords:** Recommended system · Behavior mining · Feature-cross ·
Sorting algorithm · Recommendation model

## 1 Introduction

Recommendation systems [1] have been widely used on the Internet in recent
years. Recommend suitable products for users on e-commerce websites and
increase the order rate of users. Provide news or videos that users are interested
in news and video websites to improve user stickiness. The recommendation
system is a relatively complex computer system, involving many aspects of technology. From the algorithm level alone, it can be divided into recall algorithm,

sorting algorithm, and rearrangement algorithm. This paper mainly introduces the algorithm improvement in the refinement sorting stage.

Recommendation system is the upper application of data mining. It is based on the user's historical data information to find the item that the user may click or purchase next time. The feature cross algorithm in the recommendation system ranking model can be roughly divided into three stages according to the degree of automation of the feature cross. The first stage is the manual intersection stage, which manually constructs features such as users, items, and scenes, and then manually intersects various features, such as early logistics [2]. The second stage uses semi-automatic feature crossover, such as the Xgboost+LR [3] model. Gradient boosted tree for feature intersection, final input to linear model. Finally, there is automatic feature intersection, in which there are not only linear feature intersection algorithms such as FM [4]. There are also feature intersection methods based on neural networks, such as deepFM [5], xdeepfm [6], PNN [7], CAN [8], etc.

There are many mining methods based on user historical behavior in recommender systems. It is common to calculate click-through rates, purchase rates, clicks, and purchases for various products and product categories within a time slice. Such methods have weak real-time performance and cannot capture the sequential characteristics of user behavior. So there are behavior sequence models such as DIN [9] and DIEN [10].

This paper mainly combines the method of automatic feature intersection and the method of user behavior sequence analysis. And improve CTR with different product vector representations.

The main contributions of this paper include three aspects:

1) We propose an algorithm framework to mine the user's behavior sequence information while mining the intersection of multi-dimensional features such as users, items, and scenes in the recommender system.
2) In terms of item initialization vector representation, we tried different supervised and unsupervised representation methods, and finally found that the semantic representation has a certain improvement.
3) When using a Bert [11] pretrained model as the semantic representation of news items, it is better to connect an MLP layer at the end instead of using Bert CLS directly.

## 2 Related Work

In this section, we discuss existing feature cross-mining and user behavior sequence modeling respectively.

### 2.1 Feature Cross Mining

Feature cross-mining plays an important role not only in recommender systems, but also in overall machine learning tasks. Traditional cross-feature mining often relies on manual prior knowledge, which is then calculated by statistical algorithms. This method requires a lot of prior knowledge and business experience, and it is too cumbersome to manually extract cross-features. Later, tree models are used to learn the intersection between different features, usually the XGBoost+LR [3] method in recommender systems. The benefit of this approach is that it reduces people's participation in the feature intersection process, reduces people's understanding of business knowledge, and greatly reduces people's workload. However, this method adopts a two-stage online method, which will greatly reduce the efficiency of model updating in practical applications, resulting in too cumbersome online process.

In the era of deep learning, neural networks have their own feature crossover function, but this feature crossover function is weak. If the number of network layers is too large, the noise of the features during the crossover process will also be amplified, which will have certain adverse effects. And in the crossover process, the cross-interpretability of all features is weak.

On the basis of neural network, many feature cross neural network models have been developed. We can classify models into meaningful feature intersection models and meaningless feature intersection models according to whether the feature intersection has practical significance. We will choose a part of the model, or a certain structure in the model, as the feature intersection network of our algorithm.

## 2.2  User Behavior Sequence Mining

User behavior features are crucial for recommender system ranking models. The user's historical behavior can reveal the user's hobbies. Mining user preference information plays an important role.

Traditional user historical behavior data mining usually adopts statistical methods, and counts users' purchase clicks and other behaviors through different time slices, calculates the user's preference score for items, categories, and brands, and finally puts them into the model through various feature engineering. This method requires a lot of feature engineering, requires people to have a certain understanding of the business, and is not scalable.

With the application of sequence modeling and neural networks in natural language processing, we can also learn from the processing methods of recurrent neural networks in natural language. Because the user's behavior sequence can be a sentence composed of special words such as item ID. User behavior sequences are encoded by models such as RNN [12] or LSTM [13], which play the role of mining user behavior sequences.

With the recent popularity of attention mechanisms, user behavior sequences can be mined according to the ordering of items in the user behavior sequence and the attention weights of the items, such as the deep interest network DIN [9], etc.

# 3   Our Approach

This section mainly introduces the model structure and modeling steps of Feature Cross Interest Network FCI, involving feature mining, model input and output.

## 3.1   Problem Statement

Given a user $u$ and a set of items

$$V_u = \{v_1, v_2, \ldots, v_n\}$$
$$u \in U$$
$$v \in V \tag{1}$$

where $u$ is a particular user and $v$ is an item. $V$ is the set of all candidate items. $V_u$ is the set of products recalled by user $u$, and the set of products to be sorted by user $u$. The goal of our algorithm is to predict the probability value of $V_u'$ that user $u$ may click in candidate target $V_u$.

Written in the form of a function as follows:

$$\hat{y}_i = f(u, v_i) \tag{2}$$

In this article, item refers to news, unless otherwise noted, item refers to a news entity.

## 3.2   Input and Output

Our method mainly involves two parts of data input, one is the feature data traditionally used for feature interaction. Mainly through data mining to extract basic user information, item information, scene information and so on. Such as user age group, frequent login address, frequent login time period, frequent click category topK and so no. Item features such as item category, author, source, keyword, entity word, etc. cate features. Continuous features such as CTR, impressions, clicks, categorical CTR, clicks, etc. And various click-through rate characteristics of users based on item statistics, etc. Among them, categorical features are used for explicit feature intersection, and continuous features are used for implicit DNN intersection. We express in the following mathematical form:

$$x_i = \{x_i^1, x_i^2, x_i^3, \ldots x_i^n\} \tag{3}$$

The other part is the user behavior sequence. The user's behavior sequence can be represented by the following mathematical expression:

$$s = \{i_1, i_2, i_3, \ldots i_n\} \tag{4}$$

where $i$ means the user clicked on the item. In this approach, try to use different embedding layers to represent items. The first is that we use random initialization to train item embedding. The second way uses item2vec [14] to initialize and train item embedding. However, in news scenarios where there are many new items on the shelf, item embeddings are not well represented. Third, we use Bert embedded items to express certain linguistic information, even newly listed items are well represented.

The output is the score of possible clicks by user $u$ on item $v$.

## 3.3  Model Structure

In this section, we will introduce the specific model structure of the FCI method, including feature intersection network, item Embedding layer, user interest extraction layer, prediction layer, etc. The specific structure of the model is shown in Fig. 1.

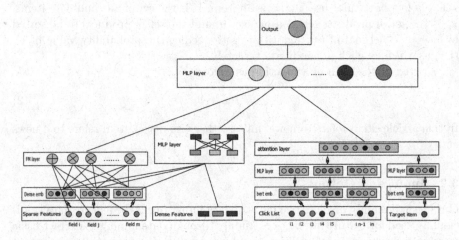

**Fig. 1.** Model structure

As shown in Fig. 1, the model mainly has three tower structures, which are mainly used to process different data. On the far left is the discrete feature processing and feature intersection module, which mainly includes the embedding layer and the feature cross layer. The intermediate structure is the implicit cross layer, which mainly uses a full-link neural network to process the embedded discrete and continuous features. On the far right is the user behavior sequence mining module, which is mainly used to process user behavior sequences and target items. It mainly includes Bert embedding layer, MLP layer and attention layer to mine the relationship between the current product and the user's historical behavior. Finally, the three parts are combined through a fully connected layer.

The combination of discrete feature intersection module and implicit intersection module is the feature intersection module. In this paper, this module adopts the deepFM model, and can also be replaced with other feature intersection models such as FWFM [15], DCN [16,17], FMFM [18], and PNN [7] according to the needs of the scene.

### 3.3.1 Sparse Feature Cross Module

This part of the model is mainly divided into two layers. The first layer is the Embedding layer, which uses direct mapping to map features to learnable vectors. Dense emb is the embedding vector result of Sparse Features.

The second layer is the feature cross network. In this paper, we use the FM model for the cross of sparse features. FM consists of the intersection of a first-order linear model and second-order features. The specific mathematical expression is as follows:

$$y = w_0 + \sum_{i=1}^{n} w_i x_i + \sum_{i=1}^{n} \sum_{j \geq i}^{n} w_{ij} x_i x_j \tag{5}$$

The first term on the right side of the equation is the zero-order bias, the second term is the first-order feature, which is mainly used to learn the impact of individual features on the model results, and the third term is the feature cross term used to learn the impact of cross features on the model results. It can be seen from the formula that the amount of calculation is large in the process of feature cross calculation, and the training of the parameter $W_{ij}$ is more difficult. Thus, the authors of FM can compute the cross term by subtracting the sum of the squares from the sum of the sums, simplifying and reducing the computational complexity. The specific mathematical expression is as follows:

$$y = w0 + \sum_{i=1}^{n} w_i x_i + 1/2 \sum_{l=1}^{k} [(\sum_{i=1}^{n} v_{i,l} x_i)^2 - \sum_{i=1}^{n} v_{i,l}^2 x_i^2] \tag{6}$$

### 3.3.2 Implicit Cross Module

This module adopts a multi-layer full-link neural network design, adopts the DNN design module in deepFM, and shares the embedding vector of the same discretized feature with the FM layer. DNN can also learn the ability of feature crossover, but the feature crossover learned by DNN contains some noise, and the crossover ability is weak. Dense features are continuous features received by the model.

### 3.3.3 User Behavior Sequence Extraction Module

This module is mainly composed of a three-layer network structure, including the embedding layer that Bert encodes the item, the MLP layer that whitens the embedding vector, and the attention layer that extracts user interests.

The BERT embedding layer based on item title encoding can effectively solve the cold start problem of newly listed items caused by traditional ID encoding. Throughout the experimental phase, we tried three different ID embedding schemes. The first is to map and encode the ID directly. Due to the large number of item and the lack of learning, this approach is generally effective. The second: The encoding method adopts the same encoding method as the first one, but the item vector pre-trained by the item2Vec method can effectively solve the shortcoming of insufficient training due to the large number of items, but this method is not suitable for newly started projects. Or the long tail is not friendly enough. The third : In order to better solve the impact of new item and long-tail items on the model, we use coding from the perspective of content coding. In the field of NLP, Bert came naturally into our field as an efficient pre-trained model. The specific mathematical expression is as follows:

$$x_i = BERT([w_1, \ldots, w_n]) \tag{7}$$

where $x_i$ represents a vector from Bert. This paper uses the CLS of the last layer of Bert to represent the vector of the item, and $[w_1, \ldots, w_n]$ represent the word of the item title.

MLP whitening layer: The vector dimension calculated by Bert is very high, reaching 768 dimensions, and the discriminative power of Bert CLS vector is not good. We employ an MLP layer to reduce the dimensionality of the Bert vector and improve its recognition ability.

Attention Layer: In order to mine the importance of the item in the user's historical behavior sequence to the current item to be sorted, an attention layer is added. The first suggestion for this approach came from DIN [9]. In this paper, several different attention methods are compared. Finally, it is found that the indicators of different attention methods are not very different in our scene. Finally, a simpler way to compute the dot product is chosen to compute the attention weights. The specific weight calculation formula and vector merging formula are as follows:

$$s(q, k) = q^T k$$
$$x = \sum_{i=1}^{n} s_i x_i \tag{8}$$

Among them, $q$ represents the target item, $k$ is an item in the user behavior sequence, and $x_i$ is the vector of items.

### 3.4  Output Layer

The design of the output layer is mainly to obtain the click probability of the user on the item. This layer concatenates the vectors of the feature intersection module and the user behavior sequence mining module, and then goes through a multi-layer MLP network, and finally uses softmax to predict the specific mathematical expression of the classification probability. The formula is as follows:

$$y = softmax(MLP([concat(x_f, x_d, x_h)])) \tag{9}$$

where $x_f$, $x_d$ and $x_h$ represent the features of explicit crossover, implicit crossover and user behavior sequence modeling, respectively.

### 3.5   Complexity Analysis

In this section, we introduce the time complexity and space complexity of each module respectively.

1. Sparse Feature Cross module
   From Eq. (6), the time complexity of the Sparse Feature Cross module can be obtained as $Time_1 = O(kn)$ and the space complexity is $Space_1 = O(1 + n + kn)$. $n$ is the number of sparse features. $k$ is the dimension of the sparse feature embedding vector.
2. Implicit Cross module
   Because the implicit cross module is a fully linked neural network. So the time complexity is $Time_2 = O(\sum_{i=1}^{n} din_i * dout_i)$ and $Space_2 = O(\sum_{i=1}^{n} din_i * dout_i)$. $n$ is the number of neural network layers. $din$ is the neural network input dimension. $dout$ is the neural network output dimension.
3. User behavior sequence extraction module
   Because this module is a combination of BERT, MLP and Attention. So the time complexity is $Time_3 = O(n^2 * d + d_{in} * d_{out} + n * d_{out})$ and $Space_3 = O(12 * (2 * n * h * in + d_{in} * d_{out} + n))$. $n$ is the length of the sequence. $d$ is the dimension of the embedding. $d_{in}$ is the input dimension of the MLP. $d_{out}$ is the dimension of the MLP output vector. $h$ is the hidden size of bert. $in$ is the intermediate size of bert.
4. Output Layer
   Output Layer is the MLP link softmax. So the time complexity is $Time_4 = O(\sum_{i=1}^{n} din_i * dout_i + v)$ and $Space_4 = O(\sum_{i=1}^{n} din_i * dout_i)$. $n$ is the number of neural network layers. $din$ is the neural network input dimension. $dout$ is the neural network output dimension. $v$ is the number of label categories
   At last
   $$Time_{all} = O(Time_1 + Time_2 + Time_3 + Time_4)$$
   $$Space_{all} = O(Space_1 + Space_2 + Space_3 + Space_4)$$

## 4   Experiment

This section will introduce the experimental related content in detail, including experimental data, experimental results, experimental comparison, and ablation experiments.

### 4.1   Data Set

In this experiment, we adopt the private feed streaming dataset and the public MIND [19] dataset. Below we will introduce the two datasets separately.

The feed stream dataset comes from the app's homepage recommendations. The data are mainly English-language financial news reports. Compared to the public dataset, this dataset is more complex and contains more long-tail data. Due to the complexity of the data, extensive feature engineering is required. For the actual data model, online CTR comparison can also be performed.

The sample labels of this dataset come from the real click data of users, and the dataset contains rich user basic data, item data and user behavior data. The dataset contains 500W pieces of sample data, positive samples are user click data, and negative samples are unclicked data of exposure.

The full name of the MIND [19] dataset is MIND: A Large-scale Dataset for News Recommendation. The data is a large-scale news recommendation dataset built by Microsoft. Data comes from real data from Microsoft News. MIND contains about 160k English news articles and more than 15 million impression logs generated by 1 million users. Every news article contains rich textual content including title, abstract, body, category and entities [19]. We sample on this dataset. The final result samples 800W samples, positive samples are news clicked by users, and negative samples are news exposed but not clicked.

After introducing the dataset, we will introduce the input feature dimension on the dataset. On the feed dataset, the dimension of sparse features is 12, the dimension of continuous features is 25, and the length of user behavior sequences is 256. On the MIND dataset, the dimension of sparse features is 9, the dimension of continuous features is 12, and the length of user behavior sequences is 512.

## 4.2   Evaluation Metrics

In this article, we will use AUC and MAP as offline metrics for the model. AUC can better represent the classification ability of the model. The actual meaning of AUC can be understood as the proportion of positive samples before negative samples in all sample pairs. The recommendation system ranking model pays more attention to the ranking results, and AUC is just suitable for this feature. As an evaluation method to measure the difference between the ideal position and the actual position in the ranking, MAP can better reflect the effect of the ranking model. In the online AB test, we use the CTR metric as an evaluation metric.

## 4.3   Model Training Details

This article uses Cross Entropy Loss as the classification loss, Bert pre-training model uses bert-base-uncased to initialize the item encoding model, the optimization function uses Adam, and learning rate is $2e-5$. This experiment uses Tesla V100 GPUs to train the model.

This experiment divides the training set, validation set and test set data according to the time dimension. According to the exposure time range of all samples, the dataset is chronologically divided into three parts of 6:2:2. The first is training data, the second is validation data, and the last is testing data.

## 4.4  Hyperparameter Settings

This section mainly introduces some hyperparameters of the algorithm and the settings of these parameters. In Sect. 4.6, we describe the parameter selection experiments in detail. The model parameters of the algorithm are mainly divided into three parts.

1. User behavior sequence length.
2. Embedding dimension of sparse features.
3. The number of MLP neurons in each module.

Next, we will explain the three types of parameters and their settings separately. The longer the user behavior sequence entered into the model, the better the long-term interest of the user can be obtained, but an excessively long sequence will lead to computational complexity. The algorithm determines the sequence length of the input algorithm based on the average behavior sequence length of users in the dataset. Therefore, the behavior length is chosen to be 256 on the Feed dataset and 512 on the MIND dataset.

The choice of embedding dimension of sparse feature is based on the ablation experiments in this paper. See 3.6 for detailed experimental details. Finally, 8 were selected on the feed dataset and 32 were selected on the MIND dataset.

The algorithm has multiple MLP layers, and the number of neurons is selected according to the ablation experiment. See 3.6 for detailed experimental details. The final results are as follows: 1. The MLP layer in the privacy intersection module adopts a two-layer full-link neural network, and the number of neurons is selected as [128, 64]. 2. The MLP layer in the user behavior extraction module adopts a single-layer full-link neural network with 64 neurons 3. The MLP layer of the output layer adopts a two-layer full-link neural network, and the number of neurons is [128, 32].

## 4.5  Experimental Comparison

This section will show a comparison of the results of different algorithms on the public dataset mind and the private feed stream recommendation dataset. The main algorithms compared are neural networks based on feature intersection, such as FM [4] and Deep FM [5]. And mainstream models such as neural network DIN [9], DIEN [10], BST [20] based on user behavior modeling. For fair comparison, the sparse and dense features are the same for all methods. DIN, DIEN, BST, and FCI all apply the same sequence of user click behaviors. The results of different methods on the dataset are shown in Table 1.

**Table 1.** The overall performance of different methods

| Method | Feed Date[a] | | | MIND Date[b] | | |
|---|---|---|---|---|---|---|
| | AUC | MAP | CTR | AUC | NDCG5 | NDCG10 |
| FM | 0.6212 | 0.1214 | 0.0271 | 0.6875 | 0.3654 | 0.4212 |
| DeepFM | 0.6352 | 0.1301 | 0.0301 | 0.6903 | 0.3701 | 0.4301 |
| DIN | 0.6389 | 0.1286 | 0.0389 | 0.6987 | 0.3787 | 0.4357 |
| DIEN | 0.6458 | 0.1251 | 0.0472 | 0.7045 | 0.3822 | 0.4482 |
| BST | 0.6571 | 0.1423 | 0.0441 | 0.7062 | 0.3911 | 0.4478 |
| FCI | **0.6881** | **0.1502** | **0.0527** | **0.7136** | **0.3957** | **0.4541** |
| Improve | 4.71% | 5.55% | 11.65% | 1% | 1.17% | 1.31% |

[a]Private News Recommendation Feed Streaming Dataset.
[b]MIcrosoft News Dataset.

As can be seen from the experimental results, both on the public dataset MIND and on the private dataset. Our algorithms have steadily improved. Our algorithm compares the feature intersection class algorithm and the user behavior sequence modeling class algorithm. The hypothesis of our algorithm is verified, effectively combining feature intersection and user behavior sequence modeling. The improvement on the MIND dataset is lower, mainly due to the lack of item features on the MIND dataset and the lack of scene features when users click.

The indicators of the algorithm on the training set and test set can also reflect the learning ability of the algorithm. The effects of different algorithms on the training set and test set are shown in Fig. 2. It can be seen that FIC has the best effect on the training set and test set. It also proves that our algorithm can learn user intent better than pure feature intersection and user sequence extraction models. Our algorithm can combine the advantages of feature intersection and user sequence mining.

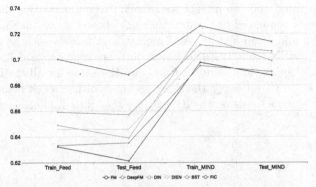

**Fig. 2.** Comparison of AUC on different algorithms

## 4.6   Ablation Experiment

This section mainly introduces the comparative ablation experiments of the algorithm between item ID and hyperparameters.

The impact of different item embedding vector methods and vector initialization on the model, the main ablation experiments are as follows.

1. Randomly initialize the item ID vector.
2. Use item2vec to initialize the item ID vector.
3. Use the Bert embedding term vector.

The specific experimental comparison is shown in Table 2:

**Table 2.** The overall performance of different methods

| Embedding method | Feed Date[a] | | | MIND Date[b] | | |
|---|---|---|---|---|---|---|
| | AUC | MAP | CTR | AUC | NDCG5 | NDCG10 |
| Randomly initialize | 0.6671 | 0.1324 | 0.0432 | 0.6925 | 0.3724 | 0.4252 |
| Item2vec | 0.6790 | 0.1455 | 0.0483 | 0.7096 | 0.3881 | 0.4311 |
| Bert | **0.6881** | **0.1502** | **0.0527** | **0.7136** | **0.3957** | **0.4541** |

[a]Private News Recommendation Feed Streaming Dataset.
[b]MIcrosoft News Dataset.

Ablation experiments between hyperparameters. This experiment involves two ablation experiments. The first is an experiment between the embedding dimensions of sparse features. The second is a comparative experiment of the number of network layers and the number of neurons in each module.

If the embedding dimension of sparse features is too low, the learned information will be weak, and if the dimension is too high, it will easily lead to overfitting. In this paper, the embedding dimension of sparse features is determined by ablation experiments. The specific experimental results are shown in Fig. 3 below.

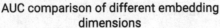

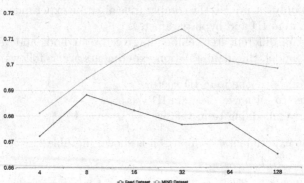

**Fig. 3.** AUC comparison of different embedding dimensions

On the Feed dataset, AUC reaches its maximum value in the 8th dimension. On the MIND dataset, AUC reaches its maximum value in the 32nd dimension. This is because MIND has a larger amount of data than feed and requires higher dimensions to fully understand the information.

The MLP layers in each module use grid search to find the optimal number of network layers and neurons. In this experiment, the Feed Stream dataset is used for the experiments. The search value of the MLP layers of the implicit intersection module, the behavior sequence extraction module, and the output module is $\{1, 2, 3\}$, and the search parameter value of the number of neurons in each layer is $\{16, 32, 64, 128, 256\}$. The final experimental results are shown in the table below.

**Table 3.** The overall performance of different ID embedding

| Model | Number of layers | Number of neurons |
|---|---|---|
| Implicit cross | 2 | [128,64] |
| User behavior mining | 1 | 64 |
| Output | 2 | [128,32] |

## 5    Conclusion

This paper proposes a new recommendation system ranking structure FCI, which combines the model framework of feature cross mining and user behavior sequence mining, and compares the impact of different item embedding methods on user behavior sequence mining. And a certain improvement has been achieved on the actual Internet industrial dataset.

In the future, we will mainly focus on the following aspects:

1. Try different feature intersection models.
2. Improve the way of mining user behavior sequences.
3. Add different data to encode the embedded item.

# References

1. Ansari, A., Essegaier, S., Kohli, R.: Internet recommendation systems. J. Mark. Res. **37**(3), 363–375 (2000)
2. Wright, R.E.: Logistic regression. J. Mol. Med (1995)
3. He, X., et al.: Practical lessons from predicting clicks on ads at Facebook. In: Proceedings of the Eighth International Workshop on Data Mining for Online Advertising, pp. 1–9 (2014)
4. Rendle, S.: Factorization machines. In: 2010 IEEE International Conference on Data Mining, pp. 995–1000 (2010)
5. Guo, H., Tang, R., Ye, Y., Li, Z., He, X.: DeepFM: a factorization-machine based neural network for CTR prediction. arXiv preprint arXiv:1703.04247 (2017)
6. Lian, J., Zhou, X., Zhang, F., Chen, Z., Xie, X., Sun, G.: xDeepFM: combining explicit and implicit feature interactions for recommender systems. In: Proceedings of the 24th ACM SIGKDD International Conference on Knowledge Discovery & Data Mining, pp. 1754–1763 (2018)
7. Qu, Y., et al.: Product-based neural networks for user response prediction. In: 2016 IEEE 16th International Conference on Data Mining (ICDM), pp. 1149–1154 (2016)
8. Bian, W., et al.: CAN: feature co-action network for click-through rate prediction. In: Proceedings of the Fifteenth ACM International Conference on Web Search and Data Mining, pp. 57–65 (2022)
9. Zhou, G., et al.: Deep interest network for click-through rate prediction. In: Proceedings of the 24th ACM SIGKDD International Conference on Knowledge Discovery & Data Mining, pp. 1059–1068 (2018)
10. Zhou, G., et al.: Deep interest evolution network for click-through rate prediction. In: Proceedings of the AAAI Conference on Artificial Intelligence, vol. 33, pp. 5941–5948 (2019)
11. Devlin, J., Chang, M.-W., Lee, K., Toutanova, K.: BERT: pre-training of deep bidirectional transformers for language understanding. arXiv preprint arXiv:1810.04805 (2018)
12. Cho, K., et al.: Learning phrase representations using RNN encoder-decoder for statistical machine translation. arXiv preprint arXiv:1406.1078 (2014)
13. Graves, A., Schmidhuber, J.: Framewise phoneme classification with bidirectional LSTM and other neural network architectures. Neural Netw. **18**(5–6), 602–610 (2005)
14. Barkan, O., Koenigstein, N.: ITEM2VEC: neural item embedding for collaborative filtering. In: 2016 IEEE 26th International Workshop on Machine Learning for Signal Processing (MLSP), pp. 1–6 (2016)
15. Pan, J., et al.: Field-weighted factorization machines for click-through rate prediction in display advertising. In: Proceedings of the 2018 World Wide Web Conference, pp. 1349–1357 (2018)
16. Wang, R., Fu, B., Fu, G., Wang, M.: Deep & cross network for ad click predictions. In: Proceedings of the ADKDD 2017, pp. 1–7 (2017)

17. Wang, R., et al.: DCN V2: improved deep & cross network and practical lessons for web-scale learning to rank systems. In: Proceedings of the Web Conference 2021, pp. 1785–1797 (2021)
18. Sun, Y., Pan, J., Zhang, A., Flores, A.: FM2: field-matrixed factorization machines for recommender systems. In: Proceedings of the Web Conference 2021, pp. 2828–2837 (2021)
19. Wu, F., et al.: MIND: a large-scale dataset for news recommendation. In: Proceedings of the 58th Annual Meeting of the Association for Computational Linguistics, pp. 3597–3606 (2020)
20. Chen, Q., Zhao, H., Li, W., Huang, P., Ou, W.: Behavior sequence transformer for e-commerce recommendation in Alibaba (2019)

# Author Index

Printed in the United States
by Baker & Taylor Publisher Services